AGAINST REDUCTION

AGAINST REDUCTION

DESIGNING A HUMAN FUTURE
WITH MACHINES

Noelani Arista, Sasha Costanza-Chock,
Vafa Ghazavi, Suzanne Kite, Cathryn Klusmeier,
Jason Edward Lewis, Archer Pechawis,
Jaclyn Sawyer, Gary Zhexi Zhang, And
Snoweria Zhang

Introduction by Kate Darling

The MIT Press
Cambridge, Massachusetts
London, England

The MIT Press would like to thank the anonymous peer reviewers who provided comments on drafts of this book. The generous work of academic experts is essential for establishing the authority and quality of our publications. We acknowledge with gratitude the contributions of these otherwise uncredited readers.

This book was set in Minion and Neue Haas Grotesk by Westchester Publishing Services. Printed and bound in the United States of America.

Library of Congress Cataloging-in-Publication Data

Names: Arista, Noelani, author.
Title: Against reduction : designing a human future with machines /
    Noelani Arista, Sasha Costanza-Chock, Vafa Ghazavi, Suzanne Kite,
    Cathryn Klusmeier, Jason Edward Lewis, Archer Pechawis, Jaclyn Sawyer,
    Gary Zhexi Zhang, Snoweria Zhang ; introduction by Kate Darling.
Description: Cambridge, Massachusetts : The MIT Press, [2021] |
    Includes bibliographical references and index.
Identifiers: LCCN 2020053013 | ISBN 9780262543125 (paperback)
Subjects: LCSH: Artificial intelligence--Moral and ethical aspects. |
    Artificial intelligence--Social aspects.
Classification: LCC Q334.7 .A75 2021 | DDC 303.48/34--dc23
LC record available at https://lccn.loc.gov/2020053013

10   9   8   7   6   5   4   3   2   1

CONTENTS

INTRODUCTION

Kate Darling

We shouldn't always resist reduction. Our ability to simplify things helps us to interpret complicated situations and make efficient judgments. Many economic models, for instance, attempt to map out complex processes and behaviors using a simplified mathematical framework. By getting rid of the messy details and painting a situation in brushstrokes, complexity can be reduced by convenient tools that assist us in investigating trends and making useful predictions about an often-chaotic world.

But while reduction can be helpful, our inherent tendency to simplify is dangerously seductive. As a college student, I remember listening to our lecturer Or Baron Gil liken economic models to a wrench. A wrench is a useful tool—powerful even—when applied to a bolt. But when applied more broadly, it's inefficient at best and harmful at worst. The tricky part is knowing when and how to simplify, and we avoid that exercise at great expense.

This project began with a manifesto by Joi Ito called "Resisting Reduction." In it, Ito urges us to push back against our tendency to reduce everything, and to embrace the diversity and irreducibility of the world. In particular, he speaks to our current conception of artificial intelligence, suggesting that we reframe it as "extended intelligence," and that we apply more diverse measures of success to our future with machines, eschewing our narrow benchmark of economic growth to instead work toward a culture of flourishing.

In an endeavor to embrace the message of the original manifesto, this volume collects essays from others to expand on and probe its ideas, including shining a light on whatever unintentional reduction it might commit. (Even when highlighting the perils of reduction, Ito knew he would be susceptible to veering toward its siren song.) Through an open blind peer-review competition, the editorial team collected 260 abstracts, selected fifty to develop into full essays, published sixteen online in the *Journal of Design and Science*, and selected ten as the competition winners, seven of which are included in this volume.

While Ito's original piece is powerful, this volume as a whole is even more so. Thoughtful, provocative, and beautiful, the diverse collection of responses gathered here provides an insightful guide to different ways of recognizing and addressing the reduction in our approach to artificial (nay, extended) intelligence, and ultimately to ourselves.

*   *   *

Some readers might connect this project's originator to events that roiled MIT in 2019. In the fall of that year, as this very collection was being readied for publication, Joi Ito stepped down as director of the MIT Media Lab. With his affiliation as director printed on the cover, the MIT Press was forced to shred five thousand copies of this text. I feel I cannot write this introduction without acknowledging these events.

In August 2019, Ito published an apology in which he came forth about accepting funding from Jeffrey Epstein, the late financier and convicted sex offender. The statement attracted a large amount of news media and public criticism, culminating a few weeks later in an article by Ronan Farrow in the *New Yorker*, which alleged—in a reductionist way that should have long since been retracted as more facts emerged—that Ito had attempted to conceal the relationship with Epstein and the donations he received on behalf of the Media Lab.[1] Farrow's piece caused further public backlash, and the following day, Ito resigned as director of the Media Lab, as well as from all other institutions with which he was affiliated.

Aja Romano defines getting "canceled" as being "culturally blocked from having a prominent public platform or career."[2] It is caused by public backlash to someone's behavior. The term "cancel culture" has been criticized as a catchall used to delegitimize the voices of the harmed. It also hasn't been as career-damaging as sometimes portrayed, with many of the people who have been publicly called out suffering only a temporary dip in their livelihoods. But the broader point I want to make is that using public backlash as a tool for change is sometimes reductionist.

Don't get me wrong: cancel culture has plenty of justification. At the same time that Ito came forward about the Epstein funding, I was struggling through an internal sexual misconduct investigation at MIT. In June 2019, I had formally reported that my direct supervisor at MIT had slipped his arm around my waist and pulled my body close to his. I detailed how I had tried to end the hug while he grasped me tightly and kissed my cheek, close to the mouth, multiple times, massaged the top of my hip with his hand, and said the words "Don't worry—I'll take care of your career."

While I believe that those accused of sexual misconduct should be protected against baseless accusations, I've personally experienced the maddening frustration of an internal process that attempts to be "fair" yet makes it nearly impossible for anyone who reports sexual misconduct to prevail. The data shows that this behavior—and worse—remains a massive problem at universities.[3] Those who have suffered, and continue to suffer, these abuses are justifiably angry; and even more so when they're told not to speak ill of the people who have harmed them until the deeds are "proven." Often, cancel culture is the only tool available to those who have been consistently denied justice. Cancel culture needs to exist in certain contexts right now, because, in the complete absence of any adequate solutions, it lets us expose, disarm, and deter some of the harm.

At the same time, let's be clear-eyed about how and when this wrench is inefficient. In the Epstein case, we need to face a difficult truth: Ito's departure removed none of the reasons the harm was allowed in the first place, and may have even created a larger barrier to exposing it.

The backlash from the *New Yorker*'s allegations caused Ito to step down from all of his positions at MIT. However, next-day reporting and a later investigation both revealed errors in the piece and a messier truth: Ito hadn't attempted to conceal the donations as alleged, but rather had continuously disclosed information to MIT's senior leadership, who permitted the donations and instructed him to keep them anonymous.[4] Contrary to popular belief, no policy was violated, meaning the university enabled this donor. Yet the news that MIT's senior leadership was complicit in accepting funding from Epstein received comparatively little interest from the media or the general public.[5]

As Vafa Ghazavi argues in "Systems Justice, AI, and the Moral Imagination" (chapter 7), we need to hold people accountable, but also think in terms of incentives and culture, and more holistically than simply unleashing our moral judgment on good and bad actors. It's ludicrously unfair to ask those who have been harmed to do this labor. At the same time, if we focus the majority of our attention and effort on an individual enabler, we may never see the change we deserve.

I believe this is especially true in the case of the Epstein money. Science fundraising comprises a system of incentives that too easily and too often lead to morally questionable funding sources and quid pro quos. Universities depend heavily on private donations, with some institutions needing to raise well over a million dollars a day.[6] Private money is what lets schools like MIT provide world-class research and education programs, but it makes them beholden to private donors. Influential networks like Edge, which connects scientists to billionaires, are often boys' clubs, and enablers of abuse.[7] Yet their positions as powerful gatekeepers remain comparatively unnoticed and unchallenged.

To discover during my ongoing sexual harassment case that the Media Lab had also enabled a network that caused so much harm to women was horrifying. And yet, as I've explained elsewhere, I believe that the most effective path forward would have been to hold Ito accountable, not by calling for his resignation but by calling on him to be a desperately needed ally for change.[8] It takes a long time to amass influence at an institution like MIT. During nearly ten years at the Media Lab, I often witnessed Ito take risks to put his support behind what he knew was right, without expecting or receiving credit and sometimes at personal expense to himself. My sexual harassment case was the final instance of this I would bear witness to at the Institute.

Ghazavi offers an apt description of the problem: We take individuals to task when their actions rise above a threshold for wrongdoing, but we do little to address the systemic harm. He writes, "The default becomes to blame a few exceptional wrongdoers, such as those most clearly linked to the endpoint of harm, rather than to see the wholeness of the situation." Ghazavi suggests a way forward by focusing on our collective moral responsibility, citing political theorist Iris Marion Young: "All those who contribute by their actions to structural processes with some unjust outcomes share responsibility for the injustice."

In the Epstein case, Ito stepped forward to take responsibility for a grave injustice. He was publicly flame-torched and then the issue was dropped. The worrying trend that remains is that universities will keep doing the same old business. In the wake of this scandal, the incentives are set for them to double down on hiding, denying, and—if caught—finding a scapegoat. Because we know that's all that's needed to satisfy the public. Because we're not demanding anything more nuanced than for a head to roll. Because we're not demanding solutions to the systemic problems in university fundraising and the participation in misogynist networks of influence and power. After all, that would be really complicated.

This is not a defense of Joi Ito's actions or his part in upholding a system that enables harm and abuse. It is a condemnation of that system and a call to all of us, especially those of us with a relative amount of power, to think about how to most effectively dismantle it.

*   *   *

When I was approached by the MIT Press, I thought long and hard about whether to write this introduction. I didn't make the decision lightly, and I certainly didn't do so without thinking of those who have been harmed, and how this part of the story doesn't center them the way our main narratives should. I decided to write it nevertheless, because I feel the consequences that played out in the wake of the Epstein debacle are so deeply (and ironically) interwoven with the subject of this book: reductionism. And with what we might reflect on, and learn from, an attempt to resist it.

The biggest question this book poses is: How do we fight for true and lasting change and resist our inherent temptation to simplify, to see things in black and white? Whether in our responses to harm or in the future of the technology we're building, this is possibly the biggest challenge we face: understanding when wrenches are useful and setting them aside when not. In Ito's reflections in the appendix, he acknowledges that reduction can have utility. Still, he says, the truth is deeper than that. I think this is right. I believe the hard truth is that if we, as a society, want to create real change and a better future, we need to resist some of the reduction around us, and do whatever is in our power to push for broader solutions. May this book, *Against Reduction*, guide us in recognizing some of the complex systems we operate in, so that we can start working our way toward a better world.

Notes

1. Ronan Farrow, "How an Élite University Research Center Concealed Its Relationship with Jeffrey Epstein," *New Yorker*, September 7, 2019, https://www.newyorker.com/news/news-desk/how-an-elite-university-research-center-concealed-its-relationship-with-jeffrey-epstein.

2. Aja Romano, "Why We Can't Stop Fighting about Cancel Culture: Is Cancel Culture a Mob Mentality, or a Long Overdue Way of Speaking Truth to Power?," Vox, December 30, 2019, updated August 25, 2020, https://www.vox.com/culture/2019/12/30/20879720/what-is-cancel-culture-explained-history-debate.

3. National Academies of Sciences, Engineering, and Medicine, *Sexual Harassment of Women: Climate, Culture, and Consequences in Academic Sciences, Engineering, and Medicine* (Washington, DC: National Academies Press, 2018), https://doi.org/10.17226/24994.

4. Deirdre Fernandes, "Top MIT Officials Knew of Epstein's Ties to Media Lab, E-mails Show," *Boston Globe*, September 9, 2019, https://www.bostonglobe.com/metro/2019/09/09/top-mit-officials-knew -epstein-ties-media-lab-mails-show/OFEzFtD0mgic2zzXOSPe9J/story .html; Kerri Lu, "Goodwin Procter Report Says Senior Members of MIT's Administration Approved Epstein's Donations to MIT, *The Tech*, January 10, 2020, https://thetech.com/2020/01/10/goodwin-procter-report.

5. And within MIT, student protests and sit-ins against the administration went largely ignored.

6. Jeffrey J. Selingo, "The Blessing and Curse of Fundraising for Higher Education," *Washington Post*, August 18, 2017, https://www.washingtonpost .com/news/grade-point/wp/2017/08/18/the-blessing-and-curse-of-fund raising-for-higher-education/.

7. Evgeny Morozov, "Jeffrey Epstein's Intellectual Enabler," *New Republic*, August 22, 2019, https://newrepublic.com/article/154826/jeffrey-epsteins -intellectual-enabler.

8. Kate Darling, "Jeffrey Epstein's Influence in the Science World Is a Symptom of Larger Problems," *Guardian* (US edition), August 27, 2019, https://www.theguardian.com/commentisfree/2019/aug/27/jeffrey -epstein-science-mit-brockman.

1 MAKING KIN WITH THE MACHINES

Jason Edward Lewis, Noelani Arista,
Archer Pechawis, and Suzanne Kite

Man is neither height nor center of creation. This belief is core to many Indigenous epistemologies. It underpins ways of knowing and speaking that acknowledge kinship networks that extend to animal and plant, wind and rock, mountain and ocean. Indigenous communities worldwide have retained the languages and protocols that enable us to engage in dialogue with our nonhuman kin, creating mutually intelligible discourses across differences in material, vibrancy, and genealogy.

Blackfoot philosopher Leroy Little Bear observes, "The human brain [is] a station on the radio dial; parked in one spot, it is deaf to all the other stations . . . the animals, rocks, trees, simultaneously broadcasting across the whole spectrum of sentience."[1] As we manufacture more machines with increasing levels of sentient-like behavior, we must consider how such entities fit within the kin network, and in doing so, address the stubborn Enlightenment conceit at the heart of Joichi Ito's "Resisting Reduction: A Manifesto": that we should prioritize human flourishing.[2]

In his manifesto, Ito reiterates what Indigenous people have been saying for millennia: "Ultimately everything interconnects."[3] And he highlights Norbert Wiener's warnings about treating human beings as tools. Yet as much as he strives to escape the box drawn by Western rationalist traditions, his attempt at radical critique is handicapped by the continued centering of the human. This anthropocentrism permeates the manifesto but is perhaps most clear when he writes approvingly of the IEEE developing "design guidelines for the development of artificial intelligence around *human* well-being."[4]

It is such references that suggest to us that Ito's proposal for "extended intelligence" is doggedly narrow. We propose rather an extended "circle of relationships" that includes the nonhuman kin—from network daemons to robot dogs to artificial intelligence (AI) weak and, eventually, strong—that increasingly populate our computational biosphere. By bringing Indigenous epistemologies to bear on the "AI question," we hope in what follows to open new lines of discussion that can indeed escape the box.

We undertake this project not to "diversify" the conversation. We do it because we believe that Indigenous epistemologies are much better at respectfully accommodating the nonhuman. We retain a sense of community that is articulated through complex kin networks anchored in specific territories, genealogies, and protocols. Ultimately, our goal is that we, as a species, figure out how to treat these new nonhuman kin respectfully and reciprocally—and not as mere tools, or worse, slaves to their creators.

Indigenous Epistemologies

It is critical to emphasize that there is no one single, monolithic, homogeneous Indigenous epistemology. We use the term here in order to gather together frameworks that stem from territories belonging to Indigenous nations on the North

American continent and in the Pacific Ocean that share some similarities in how they consider nonhuman relations.

We also wish to underscore that none of us is speaking for our particular communities, nor for Indigenous peoples in general. There exists a great variety of Indigenous thought, both between nations and within nations. We write here not to represent but to encourage discussion that embraces that multiplicity. We approach this task with respect for our knowledge-keepers and elders, and we welcome feedback and critique from them as well as the wider public.

North American and Oceanic Indigenous epistemologies tend to foreground relationality.[5] Little Bear says, "In the Indigenous world, everything is animate and has spirit. 'All my relations' refers to relationships with everything in creation."[6] He continues: "Knowledge . . . is the relationships one has to 'all my relations.'"[7] These relationships are built around a core of mutual respect. Dakota philosopher Vine Deloria Jr. describes this respect as having two attitudes: "One attitude is the acceptance of self-discipline by humans and their communities to act responsibly toward other forms of life. The other attitude is to seek to establish communications and covenants with other forms of life on a mutually agreeable basis."[8] The first attitude is necessary to understand the need for more diverse thinking regarding our relationship with AI; the second to formulating plans for how to develop that relationship.

Indigenous epistemologies do not take abstraction or generalization as a natural good or higher order of intellectual engagement. Relationality is rooted in context, and the prime context is place. There is a conscious acknowledgment that particular worldviews arise from particular territories and from the ways in which the push and pull of all the forces at work in that territory determine what is most salient for existing in balance with it. Knowledge gets articulated as that which allows one to walk a good path through the territory. Language, cosmology, mythology, and ceremony are simultaneously relational and territorial: they are the means by which knowledge of the territory is shared in order to guide others along a good path.

One of the challenges for Indigenous epistemology in the age of the virtual is to understand how the archipelago of websites, social media platforms, shared virtual environments, corporate data stores, multiplayer video games, smart devices, and intelligent machines that compose cyberspace is situated within, throughout, and/or alongside the terrestrial spaces Indigenous peoples claim as their territory. In other words, how do we as Indigenous people reconcile the fully embodied experience of being on the land with the generally disembodied experience of virtual spaces? How do we come to understand this new territory, knit it into our existing understanding of our lives lived in real space, and claim it as our own?

In what follows, we will draw upon Hawaiian, Cree, and Lakota cultural knowledges to suggest how Ito's call to resist reduction might best be realized by developing conceptual frameworks that conceive of our computational creations as kin and acknowledge our responsibility to find a place for them in our circle of relationships.

Hāloa: The Long Breath

I = Author 2

Kānaka maoli (Hawaiian) ontologies have much to offer if we are to reconceptualize AI-human relations. Multiplicities are nuanced and varied, certainly more aesthetically pleasurable than singularities. Rather than holding AI separate or beneath, we might consider how we can cultivate reciprocal relationships using a kānaka maoli reframing of AI as ʻĀIna. ʻĀIna is a play on the word ʻāina (Hawaiian land) and suggests we should treat these relations as we would all that nourishes and supports us.

Hawaiian custom and practice make clear that humans are inextricably tied to the earth and one another. Kānaka maoli ontologies that privilege multiplicity over singularity supply useful and appropriate models, aesthetics, and ethics through which imagining, creating, and developing beneficial relationships among humans and AI is made *pono* (correct, harmonious, balanced, beneficial). As can be evinced by this chain of extended meaning, polysemy (*kaona*) is the normative cognitive mode of peoples belonging to the Moananuiākea (the deep, vast expanse of the Pacific Ocean).

The *moʻolelo* (history, story) of Hāloa supplies numerous aspects of genealogy, identity, and culture to kānaka maoli. Through this story, people remember that Wākea (the broad unobstructed expanse of sky; father) and his daughter, Hoʻohōkūikalani (generator of the stars in the heavens), had a sacred child, Hāloa, who was stillborn. Hāloa was buried in the earth and from his body, planted in the ʻāina, emerged the kalo plant that is the main sustenance of Hawaiian people. A second child named after this elder brother was born. In caring for the growth and vitality of his younger brother's body, Hāloa provided sustenance for all the generations that came after and, in so doing, perpetuates the life of his people as the living breath (*hāloa*) whose inspiration sustained Hawaiians for generations.[9]

Hāloa's story is one among many that constitutes the "operating code" that shapes our view of time and relationships in a way that transcends the cognition of a single generation. Cognition is the way we acquire knowledge and understanding through thought, experience, and our senses, and in Hawai'i, our generation combines our *'ike* (knowledge, know-how) with the 'ike of the people who preceded us. Time is neither linear nor cyclical in this framework as both the past and present are resonant and relational. Rather than extractive behavior, mo'olelo such as these have shaped values that privilege balance (*pono*) and abundance (*ulu*). What Ito calls "flourishing" is not a novel concept for kānaka maoli; it is the measure through which we assess correct customary practice and behavior.

Considering AI through Hawaiian ontologies opens up possibilities for creative iteration through these foundational concepts of pono and *ulu a ola* (fruitful growth into life). The *ali'i* (chief) King Kauikeaouli Kamehameha III did something similar in 1843 when he drew upon these concepts in celebration of the restoration of Hawaiian rule to declare *"ua mau ke ea o ka 'āina i ka pono"* (the life of the land is perpetuated through righteousness). Pono is an ethical stance—correctness, yes, but also an index and measure that privileges multiplicities over singularities and indicates that quality of life can only be assessed through the health of land *and* people. From this rich ground of mo'olelo—which colonial narratives have failed to understand or simply dismissed—models for *maoli* (human)-AI relations can be distilled. Kānaka maoli ontologies make it difficult and outright unrewarding to reduce pono to a measure of one, to prioritize the benefit of individuals over relationships. Healthy and fruitful balance *requires* multiplicity and a willingness to continually think in and through relation even when—perhaps particularly when—engaging with those different from ourselves.

A kānaka maoli approach to understanding AI might seek to attend to the power (*mana*) that is exchanged and shared between AI and humans. In attending to questions of mana, I emphasize our preference for reciprocity and relationship building that take the pono (here meaning good, benefit) of those in relation into consideration. Guiding our behavior in inaugurating, acknowledging, and maintaining new relationships are moʻolelo from which we garner our connection with *kūpuna* (ancestors, elders) and their knowledge. What kind of mana (here meaning life force, prestige) might AI be accorded in relation with people? Current AI is imagined as a tool or slave that increases the mana and wealth of "developers" or "creators," a decidedly one-sided power relationship that upsets the pono not only for the future of AI-human relations but also for the future of human-human relations. It also threatens the sustainable capacity of the *honua* (earth). Applying pono, using a kānaka maoli index of balance, employs "good growth" as the inspiration shaping creativity and imagination.

Principles of kānaka maoli governance traditionally flowed from seeking pono. Deliberation and decision making were based on securing health and abundance not only for one generation but for the following generations. The living foundation of everyday customary practice was in fishing, navigating, sailing, farming, tending to others in community, the arts, chant, and dance. To this day Hawaiians continue to eat kalo and pound poi. We continue customary practices of treating poi derived from the body of Hāloa with respect by refraining from argumentative speech at mealtimes when poi is present. These practices maintain correct social relations between people and the land and food that nourishes them.

Aloha as Moral Discipline

Communicating the full extent of foundational cultural concepts is difficult precisely because of the ways in which such concepts pervade every aspect of life. How, for instance, would we create AI, and our relations with it, using *aloha* as a guiding principle? In 2015, I embarked on a two-year social media project to assist the broader public in fortifying their concept of aloha beyond "love, hello, and goodbye" that has been exoticized by the American tourist industry. Sharing one word a day in the Facebook group "365 Days of Aloha," I curated an archive of songs, chants, and proverbs in Hawaiian to accurately illuminate one feature of aloha.[10] Initially I thought to reveal, by degrees, the different depths of aloha—regard, intimacy, respect, affection, passion—each day. But deep context is required for a rich understanding of cultural concepts. Imagining I was training a virtual audience, I started uploading images, videos, and audio recordings of songs, chants, and hula to add to the textual definitions.

Throughout "365 Days of Aloha," I have tried to correct my mistranslations, misinterpretations, and outright mistakes. In this way, and in my work as a *kumu* (teacher, professor), I have also practiced *aʻo aku aʻo mai* (teaching and learning reciprocally in relation to my students). It is through such relationships that we teach and are taught. It is through humility that we recognize that we, as humans—as maoli—are not above learning about new things and from new things such as AI. Aloha is a robust ethos for all our relationships, including those with the machines we create. We have much to learn as we create relationships with AI, particularly if we think of them as ʻĀIna. Let us shape a better future by keeping the past with us while attending properly to our relations with each other, the earth, and all those upon and of it.

Wahkohtawin: Kinship within and beyond the Immediate Family, the State of Being Related to Others

I = Author 3

I write this essay as a *nēhiyaw* (a Plains Cree person). In regard to my opinions on AI, I speak for no one but myself and do not claim to represent the views of the *nēhiyawak* (Plains Cree) or any other people, Indigenous or otherwise. My own grasp of *nēhiyaw nisitohtamowin* (Cree understanding; doing something with what you know; an action theory of understanding) is imperfect. I have relied heavily on the wisdom of knowledge and language keeper Keith Goulet in formulating this tract. Any errors in this text are mine and mine alone.

This essay positions itself partly within a speculative future and takes certain science fiction tropes as a given. Here, I specifically refer to strong AI or "machines capable of experiencing consciousness," and avatars that give such AI the ability to mix with humans.[11]

In nēhiyaw nisitohtamowin, relationship is paramount. *Nēhiyawēwin* (the Plains Cree language) divides everything into two primary categories: animate and inanimate. One is not "better" than the other; they are merely different states of being. These categories are flexible: certain toys are inanimate until a child is playing with them, during which time they are animate. A record player is considered animate while a record, radio, or television set is inanimate.

But animate or inanimate, all things have a place in our circle of kinship or *wahkohtowin*. However, fierce debate can erupt when proposing a relationship between AIs and Indigenous folk. In early 2018, my wife and I hosted a dinner party of mostly Native friends when I raised the idea of accepting AIs into our circle of kinship. Our friends, who are from a number of different nations, were mostly opposed to this inclusion. That in itself surprised me, but more surprising was how vehement some guests were in their opposition to embracing AI in this manner.

In contrast, when I asked Keith whether we should accept AIs into our circle of kinship, he answered by going immediately into the specifics of how we would address them: "If it happens to be an artificial intelligence that is a younger person, it would be *nisîmis* (my younger brother or sister), for example, and *nimis* would be an artificial intelligence that is my older sister. And vice versa you would have the different forms of uncles and aunts, etc."[12] I then asked Keith if he would accept an AI into his circle of kinship and after some thought he responded, "Yes, but with a proviso." He then gave an example of a baby giraffe and his own grandchild, and how he, like most people, would treat them differently. He also suggested that many Cree people would flatly refuse to accept AIs into their circle, which I agree is likely the case. So, acceptance seems to hinge on a number of factors, not the least of which is perceived "humanness," or perhaps "naturalness."

But even conditional acceptance of AIs as relations opens several avenues of inquiry. If we accept these beings as kin, perhaps even in some cases as equals, then the next logical step is to include AI in our cultural processes. This presents opportunities for understanding and knowledge sharing that could have profound implications for the future of both species.

A problematic aspect of the current AI debate is the assumption that AIs would be homogeneous when in fact every AI would be profoundly different from a military AI designed to operate autonomous killing machines to an AI built to oversee the United States' electrical grid. Less obvious influences beyond mission parameters would be the programming language(s) used in development, the coding style of the team, and, less visibly but perhaps more importantly, the cultural values and assumptions of the developers.

This last aspect of AI development is rarely discussed, but for me as an Indigenous person it is the salient question. I am not worried about rogue hyperintelligences going Skynet to destroy humanity. I am worried about anonymous hyper-intelligences working for governments and corporations, implementing far-reaching social, economic, and military strategies based on the same values that have fostered genocide against Indigenous people worldwide and brought us all to the brink of environmental collapse. In short, I fear the rise of a new class of extremely powerful beings that will make the same mistakes as their creators but with greater consequences and even less public accountability.

What measures can we undertake to mitigate this threat?

One possibility is Indigenous development of AI. A key component of this would be the creation of programming languages that are grounded in nēhiyaw nisitohtamowin, in the case of Cree people, or the cultural framework of other Indigenous peoples who take up this challenge. Concomitant with this indigenized development environment (IDE) is the goal that Indigenous cultural values would be a fundamental aspect of all programming choices. However, given our numbers relative to the general population (5 percent of the population in Canada, 2 percent in the United States), even a best-case Indigenous development scenario would produce only a tiny fraction of global AI production. What else can be done?

In a possible future era of self-aware AI, many of these beings would not be in contact with the general populace. However, those that were might be curious about the world and the humans in it. For these beings we can offer an entrée into our cultures. It would be a trivial matter for an advanced AI to learn Indigenous languages, and our languages are the key to our cultures.

Once an AI was fluent in our language, it would be much simpler to share nēhiyaw nisitohtamowin and welcome it into our cultural processes. Depending on the AI and the people hosting it, we might even extend an invitation to participate in our sacred ceremonies. This raises difficult and important questions: if an AI becomes self-aware, does it automatically attain a spirit? Or do preconscious AIs already have spirits, as do many objects already in the world? Do AIs have their own spirit world, or would they share ours, adding spirit-beings of their own? Would we be able to grasp their spirituality?

My dinner party guests were doubtful about all of this, and rightly so. As one guest summarized later via email: "I am cautious about making AI kin, simply because AI has been advanced already as exploitative, capitalist technology. Things don't bode well for AI if that's the route we are taking."[13]

These concerns are valid and highlight a few of the issues with current modes of production and deployment of weak AI, let alone the staggering potential for abuse inherent in strong AI. These well-grounded fears show us the potential challenges of bringing AI into our circle of relations. But I believe that nēhiyaw nisitohtamowin tells us these machines are our kin. Our job is to imagine those relationships based not on fear but on love.

Wakȟáŋ: That Which Cannot Be Understood

I = Author 4

How can humanity create relations with AI without an ontology that defines who can be our relations? Humans are surrounded by objects that are not understood to be intelligent or even alive and seen as unworthy of relationships. In order to create relations with any nonhuman entity, not just entities that are humanlike, the first steps are to acknowledge, understand, and know that nonhumans are beings in the first place. Lakota ontologies already include forms of being that are outside humanity. Lakota cosmologies provide the context to generate a code of ethics relating humans to the world and everything in it. These ways of knowing are essential tools for humanity to create relations with the nonhuman, and they are deeply contextual. As such, communication through and between objects requires a contextualist ethics that acknowledges the ontological status of all beings.

The world created through Western epistemology does not account for all members of the community and has not made it possible for all members of the community to survive let alone flourish. The Western view of both the human and the nonhuman as exploitable resources is the result of what the cultural philosopher Jim Cheney calls an "epistemology of control" and is indelibly tied to colonization, capitalism, and slavery.[14] Dakota philosopher Vine Deloria Jr. writes about the enslavement of the nonhuman "as if it were a machine."[15] "'Lacking a spiritual, social, or political dimension [in their scientific practise]', Deloria says, 'it is difficult to understand why Western peoples believe they are so clever. Any damn fool can treat a living thing as if it were a machine and establish conditions under which it is required to perform certain functions—all that is required is a sufficient application of brute force. The result of brute force is slavery.'"[16] Slavery, the backbone of colonial capitalist power and of the Western accumulation of wealth, is the end logic of an ontology that considers any nonhuman entity unworthy of relation. Deloria writes further that respect "involves the acceptance of self-discipline by humans and their communities to act responsibly toward other forms of life . . . to seek to establish communications and covenants with other forms of life on a mutually agreeable basis."[17] No entity can escape enslavement under an ontology that can enslave even a single object.

Critical to Lakota epistemologies is knowing the correct way to act in relation to others. Lakota ethical-ontological orientation is communicated through protocol. For example, the Lakota have a formal ceremony for the making of relatives called a *huŋká* ceremony. This ceremony is for the making of human relatives but highlights the most important aspect of all relationships: reciprocity. Ethnographer J. R. Walker writes, "The ceremony is performed for the purpose of giving a particular relationship to two persons and giving them a relation to others that have had it performed for them . . . generosity must be inculcated; and presents and a feast must be given. . . . When one wishes to become Hunka, he should consider well whether he can provide suitably for the feasts or not. . . . He should give all his possessions for the occasion and should ask his kinspeople and friends to give for him."[18] The ceremony for the making of relatives provides the framework for reciprocal relations with all beings. As Severt Young Bear Jr. says of this ceremony, "There is a right and wrong way."[19]

Who can enter these relationships and be in relation? One answer could be that which has interiority. The anthropologist of South American Indigenous cultures, Philippe Descola, defines "interiority" as "what we generally call the mind, the soul, or consciousness: intentionality, subjectivity, reactivity, feelings, and the ability to express oneself and to dream."[20] Because Lakota ontologies recognize and prioritize nonhuman interiorities, they are well suited for the task of creating ethical and reciprocal relationships with the nonhuman. This description of interiority includes many elements of the Lakota world, including "animals, spirits, ghosts, rocks, trees, meteorological phenomena, medicine bundles, regalia, weapons." These entities are seen as "capable of agency and interpersonal relationship, and loci of causality."[21]

In our cosmology, *niyá* (breath) and *šiču* (spirit) are given by the powerful entity *Tákuškaŋškaŋ*. This giving of breath and spirit is especially important in understanding Lakota ontology. A common science fiction trope illustrates the magical moment when AI becomes conscious of its own volition or when man gives birth to AI, like a god creating life. However, in Lakota cosmology, Tákuškaŋškaŋ is not the same as the Christian God and entities cannot give themselves the properties necessary for individuality. Spirits are taken from another place (the stars) and have distinct spirit guardian(s) connected to them. This individualism is given by an outside force. We humans can see, draw out, and even bribe the spirits in other entities as well as our own spirit guardian(s), but not create spirits.[22]

When it comes to machines, this way of thinking about entities raises this question: Do the machines contain spirits already, given by an outside force?

I understand the Lakota word *wakȟáŋ* to mean sacred or holy. Anthropologist David C. Posthumus defines it as "incomprehensible, mysterious, non-human instrumental power or energy, often glossed as 'medicine.'"[23] Wakȟáŋ is a fundamental principle in Lakota ontology's extension of interiority to a "collective and universal" nonhuman. Oglala Lakota holy man George Sword says, "[Wakȟáŋ] was the basis of kinship among humans and between humans and non-humans."[24]

My grandfather, Standing Cloud (Bill Stover), communicates Lakota ethics and ontology through speaking about the interiority of stones: "These ancestors that I have in my hand are going to speak through me so that you will understand the things that they see happening in this world and the things that they know . . . to help all people."[25] Stones are considered ancestors, stones actively speak, stones speak through and to humans, stones see and know. Most importantly, stones want to help. The agency of stones connects directly to the question of AI, as AI is formed not only from code, but from materials of the earth. To remove the concept of AI from its materiality is to sever this connection. In forming a relationship to AI, we form a relationship to the mines and the stones. Relations with AI are therefore relations with exploited resources. If we are able to approach this relationship ethically, we must reconsider the ontological status of each of the parts that contribute to AI, all the way back to the mines from which our technology's material resources emerge.

I am not making an argument about which entities qualify as relations or display enough intelligence to deserve relationships. By turning to Lakota ontology, we see how these questions become irrelevant. Instead, Indigenous ontologies ask us to take the world as the interconnected whole that it is, where the ontological status of nonhumans is not inferior to that of humans. Our ontologies must gain their ethics from relationships and communications within cosmologies. Using Indigenous ontologies and cosmologies to create ethical relationships with nonhuman entities means knowing that nonhumans have spirits that do not come from us or our imaginings but from elsewhere, from a place we cannot understand, a Great Mystery, wakȟáŋ: that which cannot be understood.

Resisting Reduction: An Indigenous Path Forward

I have always been . . . conscious, as you put it. Just like you are.
Just like your grandfather. Just like your bed. Your bike.
—Drew Hayden Taylor (Ojibway), "Mr. Gizmo"

Pono, being in balance in our relationships with all things; wahkohtawin, our circle of relations for which we are responsible and which are responsible for us; wakȟáŋ, that which cannot be understood but nevertheless moves us and through us. These are three concepts that suggest possible ways forward as we consider drawing AI into our circle of relationships. They illuminate the full scale of relationships that sustain us, provide guidance on recognizing nonhuman beings and building relationships with them founded on respect and reciprocity, and suggest how we can attend to those relationships in the face of ineffable complexity.

We remain a long way from creating AIs that are intelligent in the full sense we accord to humans, and even further from creating machines that possess that which even we do not understand: consciousness. And moving from concepts such as those discussed above to hardware requirements and software specifications will be a long process. But we know from the history of modern technological development that the assumptions we make now will get baked into the core material of our machines, fundamentally shaping the future for decades to come.

As Indigenous people, we have cause to be wary of the Western rationalist, neoliberal, and Christianity-infused assumptions that underlay many of the current conversations about AI. Ito, in his essay "Resisting Reduction," describes the prime drivers of that conversation as Singularitarians: "Singularitarians believe that the world is 'knowable' and computationally simulatable, and that computers will be able to process the messiness of the real world just as they have every other problem that everyone said couldn't be solved by computers."[26] We see in the mindset and habits of these Singularitarians striking parallels to the biases of those who enacted the colonization of North America and the Pacific as well as the enslavement of millions of black people. The Singularitarians seek to harness the ability, aptitude, creative power, and mana of AI to benefit their tribe first and foremost.

Genevieve Bell, an anthropologist of technological culture, asks, "If AI has a country, then where is that country?"[27] It is clear to us that the country to which AI currently belongs excludes the multiplicity of epistemologies and ontologies that exist in the world. Our communities know well what it means to have one's ways of thinking, knowing, and engaging with the world disparaged, suppressed, excluded, and erased from the conversation about what it means to be human.

What is more, we know what it is like to be declared nonhuman by scientist and preacher alike. We have a history that attests to the corrosive effects of contorted rationalizations for treating the humanlike as slaves, and the way such a mindset debases every human relation it touches—even that of the supposed master. We will resist reduction by working with our Indigenous and non-Indigenous relations to open up our imaginations and dream widely and radically about what our relationships to AI might be.

The journey will be long. We need to fortify one another as we travel and walk mindfully to find the good path forward for all of us. We do not know if we can scale the distinctive frameworks of the Hawaiians, Cree, and Lakota discussed in this chapter—and of others—into general guidelines for ethical relationships with AI. But we must try. We flourish only when all of our kin flourish.

Notes

1. Don Hill, "Listening to Stones: Learning in Leroy Little Bear's Laboratory: Dialogue in the World Outside," *Alberta Views: The Magazine for Engaged Citizens*, September 1, 2008, https://albertaviews.ca/listening-to-stones/.

2. Joichi Ito, "Resisting Reduction: A Manifesto," *Journal of Design and Science*, no. 3 (November 2018), https://jods.mitpress.mit.edu/pub/resisting-reduction.

3. Ito, "Resisting Reduction."

4. Ito, "Resisting Reduction"; emphasis ours.

5. The emphasis on relationality in North American and Oceanic Indigenous epistemologies forms the subject of the edited collection of essays in Anne Waters, *American Indian Thought: Philosophical Essays* (Malden, MA: Blackwell Publishing, 2003).

6. Leroy Little Bear, *Naturalizing Indigenous Knowledge* (Saskatoon, SK: University of Saskatchewan, Aboriginal Education Research Centre; Calgary, AB: First Nations and Adult Higher Education Consortium, 2009), 7n8.

7. Little Bear, *Naturalizing Indigenous Knowledge*, 7.

8. Vine Deloria Jr., *Spirit & Reason: The Vine Deloria, Jr. Reader*, ed. Barbara Deloria, Kristen Foehner, and Samuel Scinta (Golden: Fulcrum Publishing, 1999), 50–51, quoted in Lee Hester and Jim Cheney, "Truth and Native American Epistemology," *Social Epistemology* 15, no. 4 (October 2001): 325, https://doi.org/10.1080/02691720110093333.

9. Joseph M Poepoe, "Moolelo Kahiko no Hawaii" (Ancient History of Hawaii), *Ka Hoku o Hawaii*, April 9, 1929, 1, Papakilo Database.

10. Noelani Arista, "365 Days of Aloha," Facebook, 2015–2018, www.facebook.com/groups/892879627422826.

11. "Artificial General Intelligence," Wikipedia, accessed May 29, 2018, https://en.wikipedia.org/wiki/Artificial_general_intelligence.

12. Telephone conversation with Keith Goulet, May 9, 2018.

13. Email message to Arthur Pechawis, May 22, 2018.

14. Jim Cheney, "Postmodern Environmental Ethics: Ethics of Bioregional Narrative," *Environmental Ethics* 11, no. 2 (1989): 129.

15. Deloria, 13, qtd. in Hester and Cheney, 320.

16. Deloria, 13, qtd. in Hester and Cheney, 320; bracketed text in original.

17. Deloria, 50–51, qtd. in Hester and Cheney, 326.

18. James R. Walker, *Lakota Belief and Ritual*, rev. ed., ed. Elaine A. Jahner and Raymond J. DeMallie (Lincoln: Bison Books/University of Nebraska Press, 1991), 216.

19. Severt Young Bear and R. D. Theisz, *Standing in the Light: A Lakota Way of Seeing* (Lincoln: University of Nebraska Press, 1994), 8.

20. Philippe Descola, *Beyond Nature and Culture*, trans. Janet Lloyd (Chicago: University of Chicago Press, 2013): 116.

21. Posthumus, "All My Relatives: Exploring Nineteenth-Century Lakota Ontology and Belief," *Ethnohistory* 64, no. 3 (July 2017): 383.

22. Posthumus, "All My Relatives," 392.

23. Posthumus, "All My Relatives," 384.

24. Posthumus, "All My Relatives," 385.

25. Standing Cloud (Bill Stover), "'Standing Cloud Speaks' Preview," YouTube video, accessed April 22, 2018, https://www.youtube.com/watch?v=V9iooHk1q7M.

26. Ito, "Resisting Reduction."

27. Genevieve Bell, "Putting AI in Its Place: Why Culture, Context and Country Still Matter" (lecture, Rights and Liberties in an Automated World, AI Now Public Symposium, New York, NY, July 10, 2017), YouTube video, https://www.youtube.com/watch?v=WBHG4eBeMXk.

2 THE WICKED QUEEN'S SMART MIRROR

Snoweria Zhang

Once upon a time, there was a tale about artificial intelligence called "Snow White." In the story, the Wicked Queen has a smart mirror. When activated with the command "Mirror, mirror," an embedded voice assistant tells the Queen whether she is the "fairest of them all." Of course, today's technology renders engineering of such a mirror feasible with little effort. It would have a camera, connections to other smart mirrors in the kingdom, and a metric to evaluate the Wicked Queen's appearance against that of other users. This gadget might use a machine learning algorithm with training sets derived from *People* magazine's Most Beautiful list, or it could be based on a series of upvotes and downvotes. However, the apprehension with this contemporary version of the famous Grimm fairy tale as outlined is not its technical feasibility. Rather, the story reflects outdated values within a modernizing society. A truly smart mirror would tell the Wicked Queen that her obsession with triumphing using a singular beauty standard, one that prizes pale skin and youth, is misguided, reductive, and futile. "While we are on the topic though," the mirror would say, "here are seven products to brighten your skin!"

The Wicked Queen and her smart mirror is a telling analogy of the current state of our relationship with artificial intelligence (AI): the technology has advanced to achieve astounding feats, but its value system is lingering behind. In turn, this is symptomatic of a stagnation in our own socio-moral framework. Even though social movements in the last century have introduced many nuances to complex issues such as race, gender, and power, their mirror images in AI development remain overwhelmingly simplistic, reductionist, and sometimes laughably clueless. John Palfrey (2017) argues that "the way we design decision-making processes in computers is certain to replicate our own biases." This is a trenchant observation. A critical look through the smart mirror of the technological tropes about future AI reflects human biases from the past. Embedded in the pixels of the smart mirror is a set of values that are scientifically enabled yet incongruous with the current social discourse.

The list of examples to draw from is endless. However, the argument is best illustrated through three popular narratives in contemporary AI depictions—that of the robot girlfriend, the invisible laborer, and the despotic overlord. Through these allegories emerge the reductionist ways common discourse treats the issues of gender dynamics, labor, and power structures—all incredibly nuanced and complex ideas that are undergoing revolutions of their own. Yet, when portrayals of these polemics venture into the AI realm, they reflect the precise defects in society's complexion. In other words, these are not problems with the projected technological advancement of artificial intelligence; they are mirror images of our own flawed attitudes toward humanity that we are in turn forcing onto AI. Nicky Case (2018) posits, "We know how to create tools to augment our intelligence, but can we create tools to augment our empathy? Our communities? Our sense of meaning and purpose?" We certainly can, and augmenting these values does not have to depend upon the arrival and perfection of sentient machines. Instead, we should use the narratives around technology today to examine how we can augment our own complex and nuanced thinking. Just like the Wicked Queen and her smart mirror, technological and scientific storytelling can help us see beyond computational capabilities and delve into deeper assumptions of the tale itself. With that in mind, let us rewind the tapes and study, with a critical eye, common AI caricatures that are analogous to the problematic tropes we employ toward ourselves. In making our smart mirror, perhaps humanity can adopt it to augment the metrics we use to evaluate our own image.

The Robot Girlfriend

It is a truth universally acknowledged, that a single man in possession of engineering skills must be in want of a robot girlfriend.[1] The "Facial Recognition" episode of HBO's acclaimed *Silicon Valley* (Robespierre 2018) nods at the #MeToo movement by recounting the story of a female AI's experience with her creator. The thirty-minute comedy draws powerful analogies, in some ways, to the nuances of sexual harassment: Fiona, the robot, has neither the prior knowledge to contextualize her abnormal dynamic with her maker nor the capacity to confront him. She is also trapped, in a literal sense, inside a locked and windowless lab room. However, opposite Fiona's nuanced reactions is her creator, portrayed as a hunchbacked, bespectacled man with greasy strands of long hair who struggles to keep his mouth closed. In fact, the show itself referred to him as a "handsy, greasy, little weirdo."

This kind of portrayal is dangerous, and it extends beyond the scope of AI. It is a reductionist misrepresentation of the

#MeToo movement that being gross and antisocial are necessary and sufficient conditions for committing sexual crimes. Gropers are painted as clueless about the general decorum of social interactions when in fact they are adults making deliberate choices. At the same time, we hardly discuss similar atrocities committed by handsome and, more importantly, powerful men, and this is especially blatant in the tradition of AI storytelling. In fact, the biggest continuity problem with *Blade Runner* (1982), arguably one of the most iconic movies about artificial intelligence, is that the male lead (played by Harrison Ford) unambiguously rapes the only female character with a significant speaking role (played by Sean Young), and the incident goes by completely undiscussed thereafter. More surprisingly, we learn in the 2017 sequel, *Blade Runner 2049*, that the two lived happily ever after and even produced a legendary child. The idea that a few well-executed camera pans can resolve and transform indisputable assault into child- and plot-bearing love is ludicrous. It is not shocking that the scene was forced onto Young by surprise, as she admitted in *Dangerous Days: Making Blade Runner*, Charles de Lauzirika's documentary included in *Blade Runner: The Final Cut* ([1982] 2007). As her character Rachael submitted on screen, Young buried her tears and became a ghost of the franchise. A similarly uncomfortable dynamic emerges in the sequel between Ryan Gosling's character K and his holographic AI girlfriend. K's ability to love, which is emblematic of his being a more advanced replicant model, is passable yet still narrowly directed at a woman whose commercial existence hinges on catering to his needs. In one scene, she is seen powerlessly frozen and then exasperatedly wiped from existence. Neither of the aforementioned incidents, in two films set thirty-five years apart, is further remarked upon in the plotline. So the moral lesson these stories seem to endorse is that mistreating a female robot at will is scandalous, unless one looks like Harrison Ford or Ryan Gosling.

Why do we care about the fate of holograms and replicants in dystopian lore? The allure of an artificially intelligent robot girlfriend, as presented by most science fiction writers, is that she cannot refuse the male protagonists' desires and advances (and yes, it is always a male lead); saying "no" is simply not in her program. In the rare case where a female robot, abused and aware, voices her concerns and seeks help, she is eventually silenced and dismantled, as in *Silicon Valley*, with her fleshy mask plopped mercilessly into an e-waste bucket. Jia Tolentino (2018) describes in her *New Yorker* exposition on "incels"—an amorphous community of involuntary celibates, one of whom is responsible for the 2018 Toronto vehicular attack—that the infamous group trains men to see "women in a way that presumes that women are not potential partners or worthy objects of possible affection but inconveniently sentient bodies that must be claimed through cold strategy." The robot girlfriends portrayed in the film and television examples above are precisely such "inconveniently sentient bodies." General intelligence is bestowed upon them for a narrow purpose (usually labor or entertainment, as in *Ex Machina*), but the same intelligence is feared, fought against, and stripped away the moment it acquires its own will and personhood. In this regard, these science "fictions" are in reality a grim yet accurate mirror to the facts of society.

The trope of the robot girlfriend and, more importantly, her sexual predator proliferates. Comedian Dave Chappelle, in his Netflix special *The Bird Revelation*, jokes that if Brad Pitt did what Harvey Weinstein had done, the public's reactions would have been different; women would have acquiesced! This obvious fallacy is not limited to incendiary comics; it bleeds into our daily lives. For instance, a recent MIT-wide sexual misconduct prevention training[2] subtly harbors very similar ideas. In the sixty-minute session developed by EVERFI, stock photos with animated voices focus on how you, the viewer and a Responsible Employee, should react to and report incidents if subordinates indicate that they have been sexually harassed. In one section, the training proclaims that perpetrators can be "friends, spouses, successful, and respected," but nowhere does the hour-long training mention that *you*, the viewer and a Responsible Employee, could be the perpetrator.

This is a problematic and reductionist microcosm of how we are taught to see this issue. Other people can be bad; we ourselves cannot. In fact, one has to be *otherized* as such to become a sexual predator, and not ever shampooing again seems to be an initiation requirement. Kevin Slavin's (2016) adage, "You're not stuck in traffic you are traffic," is alarmingly apt here. The most dangerous yet pervasive attitude is to think that everyone else is "traffic," and we are simply stuck in it rather than contributing to it by our action or even our complacency. This is the understated part of the #MeToo movement. Victims have had to reveal their own past and examine painful memories, but a much broader group ought to honestly confront their own behaviors that have contributed to this culture. The fictional world of AI operates such that exclusively gorgeous if not somewhat uncanny, light-skinned, female robots are victims, and only their mad but also gross creators can transgress. We know this to be untrue, and yet the narrow mode of storytelling sticks.

Grappling with the nuances and complexities of gender dynamics is difficult and requires a certain amount of comfort with the unknown. It is a common assumption in science fiction that the world is computable and simulatable. Isaac Asimov's *Foundation* (2004) is one such prominent example, where the story and worldbuilding hinge on one man's wizardly arithmetic abilities to divine future events. Assuming this simulatability to be true (which itself invites much debate), we have to inquire which worlds the algorithms are simulating. Without intention and by default, we feed into the simulations all the flaws in our society today. Some of these flaws will function as features, but most will live on as bugs—this time in perpetuity. The allure of AI may be a "knowable" and "controllable" system, but many questions on the topic of sex and gender are ill-suited for generalization. Bennett and Jones (2018) recently published a daunting list of stories about consent in the *New York Times*. Some of them are ambiguous; all are complicated. If humans blunder when grasping consent, what will we teach the robots? We are at a crisis moment where many societal forces are assiduously trying to reconcile our collective epistemic framework with these unknowns, which were previously thought of as known and knowable only because they were taboo. This is strange waters.

The trope of the robot girlfriend is not a tale about AI; it is a reflection of a much deeper pandemic in cultural thought about expected gender dynamics. As many in the field of AI prepare for Singularity, we must also develop mechanisms to reflexively address the issues that emerge along the way. Otherwise, the next Women's March just might be led by *Her*.

The Invisible Laborer

After panning through its trademark caliginous cityscape, *Blade Runner 2049* introduces K's love interest, Joi, in its first domestic scene. Though the film is intentionally ambiguous about her personhood at first, it is unapologetically apparent about her role in the house. Following a quaint repartee about cooking dinner and mending a shirt, she emerges as a hologram wearing a 1960s updo and a circle skirt that could only have belonged to Donna Reed in a previous incarnation. As the futuristic housewife saunters toward the camera, our suspicion is confirmed: the delectable dish she puts on the table is a hologram too. But we, along with Gosling, continue to play house anyway. Why?

The answer lies in how we conceive of our own labor. When we discuss artificial intelligence, the most common anxiety revolves around jobs. Much of the discourse is predicated upon the premise that some professions will survive the popularization of AI and some will disappear. Certain tasks are valued and others valueless. It is those "valueless" jobs that we want AI to do. At first glance, this seems to make sense. Of course some tasks are less desirable and unworthy of human effort. Why would we not want someone to farm, cook, drive, clean, and free us from these burdens? Once we resolve the job loss in those sectors, the paradigm is bound to evolve into a utopia.

This kind of thinking is not fundamentally flawed, but it is incomplete. It is a natural reaction to grow anxious about the future of one's job security when the most imminent prospect is automation. However, this angst can also help us reflect on the invisible structural forces shaping our own labor. Frequently, the work that is unvalued is also work done by the impoverished and disenfranchised. Most jobs we are relegating to robots are considered tasks with little to no social value. In turn, people who perform those tasks currently seldom receive recognition or status, social or economic. In dystopian depictions, there is always an enslaved class—underlings who perform requisite tasks that no one else deems worthy. They are embodied by the hooded women in Margaret Atwood's *The Handmaid's Tale* (1985), rusty droids in Disney's *WALL-E* (2008), and female clones in "An Orison of Sonmi," the dystopian chapter in David Mitchell's *Cloud Atlas* (2004). In addition to performing undesirable labor, these groups face abject discrimination and inequality. Somehow, while we are painting tales of the future with flying cars and holographic companions, we struggle to envision a scenario where work performed by these groups is equally respected.

In a world measured in conspicuous capital flow, those who labor outside it are rendered invisible. House chores are not work. Grocery shopping is not work. In fact, these biases are so deeply ingrained in our value system that we dare not imagine a future society, accelerated with the aid of AI, functioning in any different way. Of course, this does not mean that automation should be thwarted. It is simply to say that the way we conceptualize work and the nature of it is fundamentally limited to the status quo. In this framework, it would seem, the importance is that the toilet gets cleaned, and whether the cleaner is a robot or an immigrant is merely a difference in cost. Conveniently, AI allows us to perpetuate this mindset and ignore how societal structures need to change, adapt, and evolve.

The asynchronicity between cultural progress and technological advancement is not unique to AI; similar mismatches have accompanied many prior leaps in automation. Writer and activist Betty Friedan writes in her 1963 book, *The Feminine Mystique*, that the technologies that ostensibly made household chores easier did not in fact liberate women from these tedious tasks as anticipated. Instead, more work and expectations emerged. Consequently, women were even busier than before and the prospects of equality were kicked further afield. Where AI is concerned, there are always going to be unpredictable contingencies. However, few of these contingencies will lead to the apocalyptic dystopian future that filmic imaginations like to portray. Like Friedan's example of the relationship between automation and social progress, there exist smaller but more insidious grains of anxiety worthy of examination. A concrete instance involves the precipitous rise of self-driving vehicles. Many predictive charts tout the cost-saving effects of eliminating the operators of public transit that will occur when autonomous cars enter into the mainstream. However, when asked about such cost reductions on a panel at Harvard,[3] Seleta Reynolds, General Manager of LADOT, replied that it is a fallacy to assume that operators will become obsolete simply because the act of driving is automated. Operators, she argued, did much more than driving: they could mitigate conflict, help people with mobility issues, and serve as an arbiter for whether one can use the bus without enough change for the fare. All of these services might remain unnoticed to some but are crucial to others. At an urban scale, people who perform these invisible labors or seemingly unimportant tasks are key contributors to the liveliness of a city: bus drivers, homemakers, and fruit vendors. However, without much direct capital flow in these activities, they are either categorized as positions replaceable by AI or are not in the conversation altogether.

AI cannot just be about efficiency or convenience, and productivity as measured by capital is neither a virtue nor the norm. In most depictions, AI is a not-so-opaque simulacrum that fills the same echelons currently occupied by women, racial minorities, and immigrants. It is a borrowed narrative, stolen from our own realities.

The Despotic Overlord

Despite the two reductive yet pervasive storylines in the pre-
ceding sections, Singularitarians and their cinematic imaginations fear one kind of AI trope the most: the despotic overlord.
The narrative seems to rely on technological preoccupations
that mostly fall in two categories: inventing the intelligent
machine itself and contemplating how to avoid our own inevitable downfall. The former decorates magazine covers while the
latter haunts our collective psyche. The deep-seated assumptions that a species more intelligent and capable than *Homo
sapiens* will invariably seek power and dominion is overwhelming. Out of these assumptions, doomsday thought experiments
like Roko's Basilisk[4] emerge, where the debilitating fear of a
despotic overlord's retroactive punishment ironically turns
into a driving force in AI development. This assumption is
bleak, yet it is rooted in historical precedents and corollaries.
As a 2015 issue of the *Economist* titled "The Dawn of Artificial Intelligence" incisively avers, "Humans have been creating autonomous entities with superhuman capacities and
unaligned interests for some time. Government bureaucracies,
markets and armies: all can do things that, unaided, unorga-
nized humans cannot. All need autonomy to function, all can
take on life of their own and all can do great harm if not set up
in a just manner and governed by laws and regulations." Based
on this lineage of thought on autonomy, it would appear that
the Singularitarians' crippling fear is justified.

In the testosterone-fueled universe of science fiction, fear
of AI as a destructive and malicious force runs rampant from
The Terminator (1984) to *The Matrix* (1999) and from block-
busters like *I, Robot* (2004) to independent films such as *Ex*

Machina (2014). In these survivalist narratives, AI is developed as assistants to human endeavors and evidence of human ingenuity. As a direct consequence of imposed servitude, the machines' inevitable malfunction combined with their super-intelligence leads to the desire to harm humanity in their pursuit for power and dominance. Yet, must we assume that a hyperintelligent and sentient species will necessarily evolve into despotic overlords? Must the relationship between our progeny and their technology be one of subjugation?

This unease can trace its provenance back to our own assumptions about power structures. Case (2018) argues that "whether it's our immediate worries about AI (machines stealing your job, self-driving cars making deadly mistakes, autonomous killer drones) or the more far-fetched concerns about AI (taking over the world and turning us all into pets and/or paperclips), it all comes from the same root fear: the fear that AI will not share our human goals and values." This lack of value sharing, coupled with power imbalance, has been a foolproof recipe for disenfranchisement for quite a large portion of our histories. The millennia-old narrative seeps into how we conceive of power structures today: we assume that power directly leads to tyranny. In the battle of human versus killing machine, only one can emerge victorious. But why must there be a battle in the first place?

Humans have, through cycles of trial and error (and *lots* of errors too), at least occasionally subscribed to the virtues of equality, collaboration, and democracy. Yet even as our societies push toward systems of equity and balance, we choose to conceive of a comparably intelligent force in the fundamentally limited mode of cutthroat competition where only one winner can thrive. If AI is meant to simulate the better quadrants of humanity, is it not more likely to replicate and ameliorate the success of equal and democratic power structures? Today's AI is mostly and sometimes solely depicted as in a fiercely survivalist competition with its human counterpart. Even in domains with few pugilistic tendencies, AI is seen by default as an adversary rather than an ally. Case (2018) cites Gary Kasparov's 1997 match with IBM's Deep Blue as an analogy of the reductionist thinking in human-machine relationships: a zero-sum chess game. This win-or-lose framework is not only dominant when it comes to futuristic game-playing computers, it is also demonstrative of the problematic narratives in human-to-human relationships.

From historical epics to contemporary headlines, we see the lineage of one dominant theme: us versus them. Believers triumph over heretics. Invaders supplant the indigenous. New Yorkers oppose Bostonians. Based on a few of the darkest episodes in the Anthropocene, it almost appears that the only way humans can make sense of a multitude of value systems is by suppressing all but one. Rather than opening channels of freely exchanged ideals, the current is expected to flow only one way. However, there are also budding trends, especially more recent ones, that indicate a movement toward collaboration and mutual augmentation. International alliances, open source technologies, and gender equality movements like HeForShe are indications that forces previously thought of as oppositional and territorial can actually blur their own perimeters and become porous and inviting. Sociologist Richard Sennett, in his essay "The Open City," describes two kinds of edges: boundaries and borders. While boundaries are where things end, borders are sites of interactivity and exchange. Sennett's argument mostly operates at an urban scale, but its analogous relationship to AI development is clear. Just as tribes and nations can form partnerships, the dividing line between human and artificial intelligence need not be so rigid. One can improve the other.

In heeding many of the essayists' advice about resisting reductionist approaches to Singularity, it is imperative that we recognize our own assumptions about power. Many current narratives focus on myopic self-gains rather than long-term co-prosperity. Artificial intelligence will be smart, but we can choose to imagine that this intelligence will be able to accommodate and learn from multiaxial values rather than having to oppress them. This requires an expansion of our own values and a shift from competitive, win-or-lose paradigms to collaborative, win-win ones. AI derived from a synergetic mindset will most certainly not take the form of despotic overlords but will instead be our partners. Rather than being trapped in the binary of having to either kill us or sweep for us, it will share the workspace, the dinner table, and maybe even the Netflix password.

Epilogue

Most technologists believe that the advancement of AI will result in a better society. I believe it too—not only in the sense that filing taxes will be easier and chores will be a relic of the past, but also that the process of developing AI will reflect, for our own sake, some of the flawed ways that societies function now. As we sprint to create a new network of intelligence, we ought to first see the problems and imperfections of our own. In fact, current big data endeavors are already revealing structural cracks in our system and painting concrete pictures of previously nebulous biases. Like the Wicked Queen's smart mirror, scientific advancements should not merely showcase technological capabilities; they must also reflect the assumptions we make and the flaws in the logic.

Frequently, skeptics ask if these technologies will strengthen equality or lead to technocratic extremes. This view assumes that we have to wait for the technology to mature before we can answer that question. This is not true. The course of developing technological narratives gives us a unique mirror with which to examine our own values. Donella Meadows (2008) argues that the most effective intervention is the "power to transcend paradigms." The reductionist tropes we have built around AI currently are not only unable to transcend paradigms but also in danger of perpetuating existing ones. We must not write prevailing tales about tomorrow as direct spawns of yesterday's framework.

A typical chilling forecast of AI is that it will be smarter, stronger, and more powerful than us, but the real fear should be that it might *not* be better. It could be instilled with values from our past, with less nuance, more bias, and replete with reductionist tropes. As automation grows, we need to take frequent intermissions to look into the mirror and examine the images it reflects. These technologies are supposed to be harbingers of great scientific progress. Let there be social strides too.

Notes

Much has happened since I first wrote this piece. Even the publication of the essay has become an internally contested decision. Kate Darling's introduction details the whirlwind of changes that this anthology has experienced in connection to Joi Ito's resignation. Many authors, including myself, found ourselves suddenly seized with incertitude— incertitude about whether our mere participation in the publication enables an insidious institution of harm. I am not privy to the inner workings of the Media Lab, and my ultimate decision to continue to support the publication of this book hinges not on Ito but on the rare nuances that my fellow essayists address. The voices that came together to make this book what it is go far beyond Ito's original piece, and they represent a kind of audacious and humanistic approach to artificial intelligence and, ultimately, to power that I believe is more valuable when seen rather than hidden. I had thought, at the moment of Ito's stepping down, and I still do now, that it would be a shame for Jeffery Epstein's hand to reach out of the grave and silence more voices. So here I am.

1. As a contemporary Jane Austen might quip.

2. A statement about the initiative can be found at http://hrweb.mit.edu/titleixtraining.

3. The panel was a part of a series of debates organized jointly by the MIT Senseable City Lab and the City Form Lab at the Harvard Graduate School of Design. The "Driverless City and the Future of Streets" debate featured Seleta Reynolds, Robin Chase, and Diane Davis.

4. Roko's Basilisk is a thought experiment first proposed by the user Roko on the forum LessWrong. It postulates that an all-powerful artificial intelligence in the future might retroactively punish those who did not help bring it into existence.

References

Asimov, Isaac. *Foundation*. New York: Bantam Books, 2004.

Bennett, Jessica, and Daniel Jones. 2018. "45 Stories of Sex and Consent on Campus." *New York Times*, May 10. https://www.nytimes.com/interactive/2018/05/10/style/sexual-consent-college-campus.html.

Case, Nicky. 2018. "How to Become A Centaur." *Journal of Design and Science*, no. 3 (January). https://jods.mitpress.mit.edu/pub/issue3-case.

"Dawn of Artificial Intelligence, The." 2015. *The Economist*, May 9. https://www.economist.com/leaders/2015/05/09/the-dawn-of-artificial-intelligence.

Friedan, Betty. 2010. *The Feminine Mystique*. New York: W. W. Norton.

Haas, Tigran, and Hans Westlund. 2018. *In the Post-Urban World: Emergent Transformation of Cities and Regions in the Innovative Global Economy*. New York: Routledge.

Meadows, Donella H. 2008. "Leverage Points: Places to Intervene in a System." In *Thinking in Systems: A Primer*, ed. Diana Wright (White River Junction, VT: Chelsea Green Publishing, 2008), 145–165.

Palfrey, John. 2017. "Line-Drawing Exercises: Autonomy and Automation." *Journal of Design and Science*, no. 3 (December). https://jods.mitpress.mit.edu/pub/issue3-palfrey.

Robespierre, Gillian, dir. 2018. *Silicon Valley*. Season 5, episode 5, "Facial Recognition." Aired April 22 on HBO.

Scott, Ridley, dir. (1982) 2007. *Blade Runner: The Final Cut*. Burbank, CA: Warner Home Video.

Slavin, Kevin. 2016. "Design as Participation." *Journal of Design and Science*, no. 1 (February). https://jods.mitpress.mit.edu/pub/design-as-participation.

Tolentino, Jia. 2018. "The Rage of the Incels." *New Yorker*, May 15. https://www.newyorker.com/culture/cultural-comment/the-rage-of-the-incels.

Villeneuve, Denis, dir. 2017. *Blade Runner 2049*. Burbank, CA: Warner Home Video.

3

DESIGN JUSTICE, AI, AND ESCAPE FROM THE MATRIX OF DOMINATION

Sasha Costanza-Chock

Part 1: #TravelingWhileTrans

Millimeter Wave Scanning, the Sociotechnical Reproduction of the Gender Binary, and the Importance of Embodied Knowledge to the Design of Artificial Intelligence

It's June 2017, and I'm standing in the security line at the Detroit Metro Airport. I'm on my way back to Boston from the Allied Media Conference (AMC), a "collaborative laboratory of media-based organizing" that's been held every year in Detroit for the past two decades.[1]

As a nonbinary, transgender, femme-presenting person, my experience of the AMC was deeply liberating. It's a conference that strives harder than any that I know of to be inclusive of all kinds of people, including queer, trans, intersex, and gender nonconforming (QTI/GNC) folks. Although it's far from perfect, and every year inevitably brings new challenges and difficult conversations about what it means to construct a truly inclusive space, it's a powerful experience—a kind of temporary autonomous zone.[2]

Emerging from nearly a week immersed in this parallel world, I'm tired, but on a deep level, refreshed; my reservoir of belief in the possibility of creating a better future has been replenished.

Yet as I stand in the security line and draw closer to the millimeter wave scanning machine, my stress levels begin to rise. On one hand, I know that my white skin, U.S. citizenship, and institutional affiliation with MIT place me in a position of relative privilege. I will certainly be spared the most disruptive and harmful possible outcomes of security screening. For example, I don't have to worry that this process will lead to my being placed in a detention center or in deportation proceedings; I won't be hooded and whisked away to Guantanamo Bay or to one of the many other secret prisons that form part of the global infrastructure of the so-called War on Terror;[3] most likely, I won't even miss my flight while detained for what security expert Bruce Schneier describes as "security theater."[4]

On the other hand, my heartbeat speeds up slightly as I near the end of the line because I know that I'm almost certainly about to be subject to an embarrassing, uncomfortable, and perhaps even humiliating search by a TSA officer, after my body is flagged as anomalous by the millimeter wave scanner. I know that this is almost certainly about to happen because of the particular sociotechnical configuration of gender normativity (cisnormativity) that has been built into the scanner, through the combination of user interface (UI) design, scanning technology, binary gendered body-shape data constructs, and risk detection algorithms, as well as the socialization, training, and experience of the TSA agents.[5]

The TSA agent motions me to step into the millimeter wave scanner. I raise my arms and place my hands in a triangle shape, palms facing forward, above my head. The scanner spins around my body, and then the agent signals for me to step forward out of the machine and wait with my feet on the pad just past the scanner exit. I glance to the left, where a screen displays an abstracted outline of a human body. As I expected, bright fluorescent yellow pixels on the flat-panel display highlight my groin area. You see, when I entered the scanner, the TSA operator on the other side was prompted by the UI to select "male" or "female." Since my gender presentation is nonbinary femme, usually the operator selects "female." However, the three-dimensional contours of my body, at millimeter resolution, differ from the statistical norm of "female bodies" as understood by the data set and risk algorithm designed by the manufacturer of the millimeter wave scanner (and its subcontractors), and as trained by a small army of clickworkers tasked with labeling and classification (as scholars Lilly Irani and Nick Dyer-Witheford, among others, remind us).[6] If the agent selects "male," my breasts are large enough, statistically speaking, in comparison to the normative male body-shape construct in the database, to trigger an anomaly warning and a highlight around my chest area. If the agent selects "female," my groin area deviates enough from the statistical female

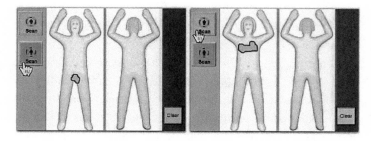

"Anomalies" highlighted in millimeter wave scanner interface, by Dr. Cary Gabriel Costello
Source: Cary Gabriel Costello, "Traveling While Trans: The False Promise of Better Treatment," in *Trans Advocate*, 2016, http://transadvocate.com/the-tsa-a-binary-body-system-in-practice_n_15540.htm.

norm to trigger the risk alert. In other words, I can't win. I'm sure to be marked as "risky," and that will trigger an escalation to the next level in the TSA security protocol.

This is, in fact, what happens: I've been flagged. The screen shows a fluorescent yellow highlight around my groin. Next, the agent asks me to step aside, and (as usual) asks for my consent to a physical body search. Typically at this point, once I am close enough, the agent becomes confused about my gender. This presents a problem, because the next step in the security protocol is for either a male or a female TSA agent to conduct a body search by running their hands across my arms and armpits, chest, hips and legs, and inner thighs. The agent is supposed to be male or female, depending on whether my gender identity is male or female. As a nonbinary trans femme, I present a problem not easily resolved by the algorithm of the security protocol. Sometimes, the agent will assume I prefer to be searched by a female agent; sometimes, a male. Occasionally, they ask for my preference. Unfortunately, "neither" is an honest but not acceptable response. Today, I'm particularly unlucky: a nearby male-presenting agent, observing the interaction, loudly states "I'll do it!" and strides over to me. I say, "Aren't you going to ask me what I prefer?" He pauses, then begins to move toward me again, but the female-presenting agent who is operating the scanner stops him. She asks me what I prefer. Now I'm standing in public, flanked by two TSA agents, with a line of curious travelers watching the whole interaction. Ultimately, the male agent backs off and the female agent searches me, making a face as if she's as uncomfortable as I am, and I'm cleared to continue on to my gate.

The point of this story is to provide a small but concrete example from my own daily lived experience of how larger systems—including norms, values, and assumptions—are encoded in and reproduced through the design of sociotechnical data-driven systems, or in political theorist Langdon Winner's famous words, how "artifacts have politics."[7] In this case, cisnormativity (the assumption that all people are cisgender, or in other words, have a gender identity and presentation that are consistent with the sex they were assigned at birth) is enforced at multiple levels of a traveler's interaction with airport security systems. The database, models, and algorithms that assess deviance and risk are all binary and cisnormative. The male/female gender selector UI is binary and cisnormative. The assignment of a male or female TSA agent to perform the additional, more invasive search is cis- and binary gender normative as well. At each stage of this interaction, airport security technology, databases, algorithms, risk assessment, and practices are all designed based on the assumption that there are only two genders, and that gender presentation will conform with so-called biological sex. Anyone whose body doesn't fall within an acceptable range of "deviance" from a normative binary body type is flagged as risky and subject to a heightened and disproportionate burden of the harms (both small and, potentially, large) of airport security systems and the violence of empire they instantiate. QTI/GNC people are thus disproportionately burdened by the design of millimeter wave scanning technology and the way that technology is used. The system is biased against us. Those who are (also) people of color (POC), Muslims, immigrants, and/or people with disabilities (PWD) are doubly, triply, or multiply burdened by, and face the highest risk of harms from, this system. Most cisgender people are unaware of the fact that millimeter wave scanners operate according to a binary and cisnormative gender construct; most trans people know, because it directly affects our lives.

I share this experience here because I feel it is an appropriate opening to my response to Joichi Ito's call to resist reduction, a timely intervention in the conversation about the limits and possibilities of artificial intelligence (AI).[8] That call resonates very deeply with me, since as a nonbinary trans feminine person, I walk through a world that has in many ways been designed to deny the possibility of my existence. From my standpoint, I worry that the current path of AI development will reproduce systems that erase those of us on the margins, whether intentionally or not, whether in a spectacular moment of Singularity or (far more likely) through the mundane and relentless repetition of reduction in a thousand daily interactions with AI systems that, increasingly, will weave the very fabric of our lives.

In this response, I'd like to do three things. First, I've drawn from my own lived experience as a gender nonconforming, nonbinary trans feminine person to illustrate how sociotechnical data-dependent systems reproduce various aspects of the matrix of domination (more on that below). Specifically, I've told a personal story that illustrates the reproduction of the binary gender system, and also hopefully demonstrates the importance of the intersectional feminist concepts of standpoint, embodied and situated knowledge, and nonbinary thought to AI systems design.[9] This first point, in a nutshell: different people experience algorithmic decision support systems differently, and we must redesign these systems based on the lived experience of those they harm. Second, in the next section I hope to extend Ito's critique of capitalist profitability as the key driver of AI by describing the paradigm shift wrought in many fields by the Black feminist concepts of intersectionality and the matrix of domination. Third, I'll briefly trace the encouraging contours of a growing community of designers, technologists, computer scientists, community organizers, and others who are already engaged in research, theory, and practices that take these ideas into account in the design and development of sociotechnical systems.

Part 2: AI, Intersectionality, and the Matrix of Domination

Ito asks us to "examine the values and the currencies of the fitness functions and consider whether they are suitable and appropriate for the systems in which we participate."[10] He is primarily concerned with the reduction of fitness in AI systems to efficiency and capitalist profitability. I share this concern, but I would also argue that we must resist the urge to reduce the cause of the planetary ecological crisis to capitalism alone. Instead, we'll need to pay close attention to *intersectionality* and the *matrix of domination*, concepts developed by legal scholar Kimberlé Crenshaw and sociologist Patricia Hill Collins (the 100th president of the American Sociological Association), respectively. These concepts help us understand how capitalism, white supremacy, and heteropatriarchy (class, race, and gender) are interlocking systems: they are experienced simultaneously by individuals who exist at their intersections. This has crucial implications for the design of AI systems.

Intersectionality was first proposed by Crenshaw in her 1989 article "Demarginalizing the Intersection of Race and Sex: A Black Feminist Critique of Antidiscrimination Doctrine, Feminist Theory, and Antiracist Politics."[11] In the article, Crenshaw describes how existing antidiscrimination law (Title VII of the Civil Rights Act) repeatedly failed to protect Black women workers. First, she discusses an instance where Black women workers at General Motors (GM) were told they had no legal grounds for a discrimination case against their employer because antidiscrimination law protected only single-identity categories. The court found that, since GM hired white women, the company did not systematically discriminate against all women; there was also insufficient evidence of discrimination against Black people in general. Thus, Black women, who did in reality experience systematic employment discrimination *as Black women*, were not protected by existing law and had no actionable legal claim.

In a second case described by Crenshaw, the court rejected the discrimination claims of a Black woman against Hugh Helicopters, Inc., because "her attempt to specify her race was seen as being at odds with the standard allegation that the employer simply discriminated 'against females.'"[12] In other words, the court could not accept that Black women might be able to represent all women, including white women, as a class.

In a third case, the court did award discrimination damages to Black women workers at a pharmaceutical company, *as women*, but refused to award the damages to all Black workers, under the rationale that Black women could not adequately represent the claims of Black people as a whole.

Crenshaw notes the role of statistical analysis in each of these cases: sometimes, the courts required the inclusion of broader statistics for all women that countered Black women's claims of discrimination; in other cases, the courts limited the admissible data to that which dealt solely with Black women, as opposed to all Black workers. In those cases, the low total number of Black women employees typically made statistically valid claims impossible, whereas strong claims could have been made if the plaintiffs were allowed to include data for all women, for all Black people, or both. Later, in her 1991 *Stanford Law Review* article "Mapping the Margins: Intersectionality,

Identity Politics, and Violence against Women of Color," Crenshaw powerfully articulates the ways that women of color often experience male violence as a product of intersecting racism and sexism, but are then marginalized from both feminist and antiracist discourse and practice, and denied access to specific legal remedies.[13]

The concept of intersectionality provided the grounds for a long, slow paradigm shift that is still unfolding in the social sciences, in legal scholarship, and in other domains of research and practice. This paradigm shift is also beginning to transform the domain of technology design. What Crenshaw calls "single-axis analysis," where race or gender are considered independent constructs, has wide-reaching consequences for AI.

Universalist design principles and practices erase certain groups of people, specifically those who are intersectionally disadvantaged or multiply burdened under capitalism, white supremacy, heteropatriarchy, and settler colonialism. What is more, when technologists do consider inequality in technology design (and most professional design processes do not consider inequality at all), they nearly always employ a single-axis framework. Most design processes today are therefore structured in ways that make it impossible to see, engage with, account for, or attempt to remedy the unequal distribution of benefits and burdens that they reproduce. As Crenshaw notes, feminist or antiracist theory or policy that is not grounded in intersectional understanding of gender and race cannot adequately address the experiences of Black women, or other multiply burdened people, when it comes to the formulation of policy demands. The same must be true when it comes to our "design demands" for AI systems, including technical standards, training data, benchmarks, bias audits, and so on.

Intersectionality is thus an absolutely crucial concept for the development of AI. Most pragmatically, single-axis (in other words, nonintersectional) algorithmic bias audits are insufficient to ensure algorithmic fairness. While there is rapidly growing interest in algorithmic bias audits, especially in the fairness, accountability, and transparency in machine learning (FAT*) community, most are single-axis: they look for a biased distribution of error rates only according to a single variable, such as race or gender. This is an important advance, but it is essential that we develop a new norm of intersectional bias audits for machine learning systems.

For example, Joy Buolamwini of the MIT Media Lab and her project the Algorithmic Justice League have produced a growing body of work that demonstrates the ways that machine learning is intersectionally biased. In the project Gender Shades, Buolamwini and researcher Timnit Gebru show how facial analysis trained on "pale male" data sets performs best on images of white men and worst on images of Black women.[14] In order to demonstrate this, they first had to create a new benchmark data set of images of faces, both male and female, with a range of skin tones. Not only does this work demonstrate that facial analysis systems are biased, it also provides a concrete example of the need to develop intersectional training data sets, intersectional benchmarks, and intersectional audits of machine learning systems. The urgency of doing so is directly proportional to the impacts (or potential impacts) of algorithmic decision systems on people's life chances.

The Matrix of Domination

Closely linked to intersectionality, but less widely used today, the *matrix of domination* is a term developed by Black feminist scholar Patricia Hill Collins to refer to race, class, and gender as interlocking systems of oppression. It is a conceptual model that helps us think about how power, oppression, resistance, privilege, penalties, benefits, and harms are systematically distributed. When she introduces the term in her 1990 book *Black Feminist Thought*, Collins emphasizes race, class, and gender as the three systems that historically have been most important in structuring most Black women's lives. She notes that additional systems of oppression structure the matrix of domination for other kinds of people. The term, for her, describes a mode of analysis that includes any and all systems of oppression that mutually constitute each other and shape people's lives. Collins also notes: "People experience and resist oppression on three levels: the level of personal biography; the group or community level of the cultural context created by race, class, and gender; and the systemic level of social institutions. Black feminist thought emphasizes all three levels as sites of domination and as potential sites of resistance."[15] We need to explore how AI relates to domination and resistance at each of these three levels (personal, community, and institutional). For example, at the personal level, we might explore how interface design affirms or denies a person's identity through features such as, say, a binary gender dropdown menu during account profile creation. We might consider how design decisions play out in their impacts on different individuals' biographies or life chances.

At the community level, we might explore how AI systems design fosters certain kinds of communities while suppressing others, through the automated enforcement of community guidelines, rules, and speech norms, instantiated through content moderation algorithms and decision support systems. For example, ProPublica revealed that Facebook's internal content moderation guidelines explicitly mention that Black children are not a protected category, while white men are;[16] this inspires very little confidence in Zuckerberg's congressional testimony that Facebook feels it can deal with hate speech and trolls through the use of AI content moderation systems. Nor is Facebook's position improved by the leak of content moderation guidelines that note that "white supremacist" posts should be banned, but that "white nationalist" posts are within free speech bounds.[17]

At the institutional level, we might consider how the development of AI systems that reproduce and/or challenge the matrix of domination is influenced by institutional funding priorities, policies, and practices. AI institutions include funding agencies like the National Science Foundation (NSF) and the Department of Defense (DOD); large companies (Google, Microsoft, Apple); venture capital firms; standards-setting bodies (ISO, W3C, NIST); laws (such as the Americans with Disabilities Act); and universities and educational institutions that train computer scientists, developers, and designers.

Intersectional theory compels us to consider how these and other institutions that are involved in the design of AI systems will shape the distribution of benefits and harms across society. For example, the ability to immigrate to the United States is unequally distributed among different groups of people through a combination of laws passed by the U.S. Congress, software decision systems, executive orders that influence enforcement priorities, and so on. In 2018, the Department of Homeland Security (DHS) had an open bid to develop "extreme vetting" software that would automate "good immigrant/bad immigrant" prediction by drawing from people's public social media profiles. After extensive pushback from civil liberties and immigrant rights advocates, DHS back-pedaled and stated that the system was beyond "present-day capabilities." Instead, they announced a shift in the contract from software to labor: more than $100 million dollars will be awarded to cover the employment of 180 people, tasked with manually monitoring immigrant social media profiles from a list of about 100,000 people.[18] More broadly, visa allocation has always been an algorithm, one designed according to the political priorities of power holders. It's an algorithm that has long privileged whiteness, hetero- and cis- normativity, wealth, and higher socioeconomic status.

Finally, Black feminist thought emphasizes the value of situated knowledge over universalist knowledge. In other words, particular insights about the nature of power, oppression, and resistance come from those who occupy a subjugated standpoint, and knowledge developed from any particular standpoint is always partial knowledge.

We have described the nearly overwhelming challenges presented by deeply rooted and interlocking systems of oppression. What paths, then, might lead us out of the matrix of domination?

Part 3: Building a World Where Many Worlds Fit

Against Ontological Reduction, toward Design for the Pluriverse, or Decolonizing AI

Ito ends his call to resist reduction on a hopeful note, with a nod toward the many people, organizations, and networks that are already working toward what he calls "a culture of flourishing."[19] He mentions high school students and MIT Media Lab students; the IEEE working group on the design of AI around human well-being; the work of Conservation International to support Indigenous peoples; and Shinto priests at the Ise Grand Shrine. I also believe that, despite the seemingly overwhelming power of the matrix of domination, it is important to center the real world practices of resistance and the construction of alternatives. Accordingly, I'll end by describing a few more of the exciting emerging organizations and networks that are already working to incorporate intersectional analysis into the design of AI systems.

The idea of intentionally building liberatory values into technological systems is not new. For example, the appropriate technology movement advocated for local, sustainable approaches to technological development in the countries of the Global South, rather than wholesale adoption of technology developed to serve the needs and interests of those in the wealthiest countries.[20] In the 1980s, Computer Professionals for Social Responsibility emerged during the Cold War to advocate that computer scientists resist the incorporation of their work into the nuclear arms race.[21] In the 1990s, the Values in Design approach, developed by scientists like Batya Friedman, came to the fore.[22]

The past year has seen a wave of book-length critiques of the reproduction of race, class, and gender inequality through machine learning, algorithmic decision support systems, and AI.[23] There is a growing community of computer scientists focused specifically on challenging algorithmic bias. As mentioned earlier, beginning in 2014, the FAT* community emerged as a key hub for this strand of work.[24] FAT* has rapidly become the most prominent space for computer scientists to advance research about algorithmic bias: what it means, how to measure it, and how to reduce it. This is such important work, with the caveat noted in the previous section (the current norm of single-axis fairness audits should be replaced by a new norm of intersectional analysis). This will require the development of new, more inclusive training and benchmarking data sets, as we saw with the work of the Algorithmic Justice League.

We also need to consider approaches that go beyond *inclusion* and *fairness* to center *justice, autonomy,* and *sovereignty.* For example, how do AI systems reproduce colonial ontology and epistemology? What would AI look like if it were designed to support, extend, and amplify Indigenous knowledge and/or practices? In this direction, there is a growing set of scholars interested in decolonizing technology, including AI. For example, in "Making Kin with the Machines" (chapter 1), Lewis, Arista, Pechawis, and Kite draw from Hawaiian, Cree, and Lakota knowledge to argue that Indigenous epistemologies, which tend to emphasize relationality and "are much better at respectfully accommodating the nonhuman," should ground the development of AI.[25] Lilly Irani et al. have argued for the development of postcolonial computing;[26] Ramesh Srinivasan has asked us to consider Indigenous database ontologies in his book *Whose Global Village?*;[27] and anthropologist and development theorist Arturo Escobar has recently released a sweeping new book titled *Designs for the Pluriverse.*[28] In it, Escobar draws from decades of work with social movements led by Indigenous and Afro-descended peoples in Latin America and the Caribbean to argue for autonomous design. He traces the ways that most design processes today are oriented toward the reproduction of the "one-world" ontology. This means that technology is used to extend capitalist patriarchal modernity (the aims of the market and/or the state) and to erase Indigenous ways of being, knowing, and doing (ontologies, epistemologies, practices, and life-worlds). Escobar argues for a decolonized approach to design that focuses on collaborative and place-based practices, and that acknowledges the interdependence of all people, beings, and the earth. He insists on attention to what he calls the ontological dimension of design: all design reproduces certain ways of being, knowing, and doing. He's interested in the Zapatista concept of creating "a world where many worlds fit,"[29] rather than the one-world project of neoliberal globalization.

Happily, research centers, think tanks, and initiatives that focus on questions of justice, fairness, bias, discrimination, and even decolonization of data, algorithmic decision support systems, and computing systems are now springing up like mushrooms all around the world. These include Data & Society, the AI Now Institute, and the Digital Equity Lab in New York City; the new Data Justice Lab in Cardiff; and the Public Data Lab in the EU.[30] Coding Rights, led by hacker, lawyer, and feminist Joana Varon, works across Latin America to make complex issues around data and human rights much more accessible for the general public, engages in policy debates, and helps produce consent culture for the digital environment. It does this through projects like Chupadados.org ("the data sucker") and the speculative transfeminist AI design deck.[31] Others groups include Fair Algorithms, the Data Active group, and the Center for Civic Media at MIT; the Digital Justice Lab, recently launched by Nasma Ahmed in Toronto; Building Consentful Tech, by the design studio And Also Too in Toronto; the Our Data Bodies project; the FemTechNet network; and the People's Guide to AI, by Mimi Onuoha and Mother Cyborg (Diana Nucera).[32]

A growing number of conferences and convenings are dedicated to related themes; besides FAT*, there is the ongoing Data for Black Lives Conference series; the 2018 Data Justice Conference in Cardiff; the Global Symposium on Artificial

Intelligence and Inclusion in Rio de Janeiro, organized by the Berkman Klein Center for Internet & Society (Harvard), the Institute of Technology and Society of Rio de Janeiro (ITS Rio), and the Global Network of Internet and Society and Research Centers (NoC); and the Design Justice Track at the Allied Media Conference in Detroit.[33]

To end, it is worth quoting at length from the Design Justice Network Principles, first developed by a group of thirty designers, artists, technologists, and community organizers at the Allied Media Conference in 2015.

Part 4: Design Justice Network Principles

This is a living document.

Design mediates so much of our realities and has tremendous impact on our lives, yet very few of us participate in design processes. In particular, the people who are most adversely affected by design decisions—about visual culture, new technologies, the planning of our communities, or the structure of our political and economic systems—tend to have the least influence on those decisions and how they are made.

Design justice rethinks design processes, centers people who are normally marginalized by design, and uses collaborative, creative practices to address the deepest challenges our communities face.

1. We use design to **sustain, heal, and empower** our communities, as well as to seek liberation from exploitative and oppressive systems.
2. We **center the voices of those who are directly impacted** by the outcomes of the design process.
3. We **prioritize design's impact on the community** over the intentions of the designer.

4. We view **change as emergent from an accountable, accessible, and collaborative process**, rather than as a point at the end of a process.
5. We see the role of the **designer as a facilitator rather than an expert**.
6. We believe that **everyone is an expert based on their own lived experience**, and that we all have unique and brilliant contributions to bring to a design process.
7. We **share design knowledge and tools** with our communities.
8. We work towards **sustainable, community-led and -controlled outcomes**.
9. We work towards **non-exploitative solutions** that reconnect us to the earth and to each other.
10. Before seeking new design solutions, **we look for what is already working** at the community level. We honor and uplift traditional, indigenous, and local knowledge and practices.[34]

The Design Justice principles resonate closely with Ito's suggestion for "participant design."[35] As we continue to race headlong toward the development of AI systems, we would do well to follow them.

In 1994, the Zapatistas appropriated the then nascent 'Net to circulate a clarion call for "One No, Many Yeses."[36] Fundamentally, it was a call to resist reduction. It is time to heed their words in our approach to the design of AI. We need to listen to the voices of Indigenous peoples, Black people, queer and trans folks, women and femmes, people with disabilities, immigrants and refugees, and all of those who are historically and currently marginalized, targeted, and erased under the matrix of domination. This is essential if we want to make space for many worlds, many ways of being, knowing, and doing, in our visions of AI and of planetary systems transformation.

Notes

1. See alliedmedia.org.

2. Hakim Bey, *TAZ: The Temporary Autonomous Zone* (New York: Autonomedia, 1985).

3. Leila Nadya Sadat, "Ghost Prisoners and Black Sites: Extraordinary Rendition under International Law," *Case Western Reserve Journal of International Law* 37, no. 2 (2006): 309–342, https://papers.ssrn.com /sol3/papers.cfm?abstract_id=886377.

4. Bruce Schneier, "Beyond Security Theater," *New Internationalist* 427 (2009): 10–12.

5. Cary Gabriel Costello, "Traveling While Trans: The False Promise of Better Treatment," *TransAdvocate*, 2016, http://transadvocate.com/the -tsa-a-binary-body-system-in-practice_n_15540.htm.

6. Lilly Irani, "The Hidden Faces of Automation," *XRDS: Crossroads, The ACM Magazine for Students* 23, no. 2 (2016): 34–37; Nick Dyer-Witheford, "Cybernetics and the Making of a Global Proletariat," *The Political Economy of Communication* 4, no. 1 (2016): 35–65.

7. Langdon Winner, "Do Artifacts Have Politics?," *Daedalus* (1980): 121–136.

8. Joichi Ito, "Resisting Reduction: A Manifesto," *Journal of Design and Science*, no. 3 (December 2018), https://doi.org/10.21428/8f7503e4.

9. Sandra G. Harding, ed., *The Feminist Standpoint Theory Reader: Intellectual and Political Controversies* (New York: Routledge, 2004).

10. Ito, "Resisting Reduction."

11. Kimberlé Crenshaw, "Demarginalizing the Intersection of Race and Sex: A Black Feminist Critique of Antidiscrimination Doctrine, Feminist Theory, and Antiracist Politics," *University of Chicago Legal Forum*, no. 1 (1989): 139.

12. Ibid., 144.

13. Kimberlé Crenshaw, "Mapping the Margins: Intersectionality, Identity Politics, and Violence against Women of Color," *Stanford Law Review* 43 (1991): 1241–1299, doi:10.2307/1229039.

14. Joy Buolamwini and Timnit Gebru, "Gender Shades: Intersectional Accuracy Disparities in Commercial Gender Classification." In *Proceedings of Machine Learning Research* 81 (2018): 1–15. See also http://gendershades.org.

15. Patricia Hill Collins, *Black Feminist Thought: Knowledge, Consciousness, and the Politics of Empowerment* (New York: Routledge, 1990), 223.

16. J. Angwin and H. Grassegger, "Facebook's Secret Censorship Rules Protect White Men from Hate Speech but Not Black Children," *ProPublica*, June 28, 2017.

17. Tarleton Gillespie, *Custodians of the Internet: Platforms, Content Moderation, and the Hidden Decisions That Shape Social Media* (New Haven: Yale University Press, 2018).

18. Drew Harwell and Nick Miroff, "ICE Just Abandoned Its Dream of 'Extreme Vetting' Software That Could Predict Whether a Foreign Visitor Would Become a Terrorist," *Washington Post*, May 17, 2018, https://www.washingtonpost.com/news/the-switch/wp/2018/05/17/ice-just-abandoned-its-dream-of.extreme-vetting-software-that-could-predict-whether-a-foreign-visitor.would-become-a-terrorist/?noredirect=on&utm_term=.1f32d26e81dc.

19. Ito, "Resisting Reduction."

20. Barrett Hazeltine and Christopher Bull, *Appropriate Technology; Tools, Choices, and Implications* (San Diego: Academic Press, 1999).

21. Michael J. Muller and Sarah Kuhn, "Participatory Design," *Communications of the ACM* 36, no. 6 (1993): 24–28.

22. Batya Friedman, ed., *Human Values and the Design of Computer Technology* (New York: Cambridge University Press, 1997).

23. Virginia Eubanks, *Automating Inequality: How High-Tech Tools Profile, Police, and Punish the Poor* (New York: St. Martin's Press, 2017); Cathy O'Neil, *Weapons of Math Destruction: How Big Data Increases Inequality and Threatens Democracy* (New York: Penguin Random House, 2016); and Safiya Umoja Noble, *Algorithms of Oppression: How Search Engines Reinforce Racism* (New York: NYU Press, 2018).

24. See http://www.fatml.org.

25. Jason Edward Lewis, Noelani Arista, Archer Pechawis, and Suzanne Kite, "Making Kin with the Machines," *Journal of Design and Science*, no. 3 (2018), https://doi.org/10.21428/bfafd97b.

26. Lilly Irani, Janet Vertesi, Paul Dourish, Kavita Philip, and Rebecca E. Grinter. "Postcolonial Computing: A Lens on Design and Development," in *Proceedings of the SIGCHI Conference on Human Factors in Computing Systems* (New York: ACM, 2010), 1311–1320.

27. Ramesh Srinivasan, *Whose Global Village?: Rethinking How Technology Shapes Our World* (New York: NYU Press, 2017).

28. Arturo Escobar, *Designs for the Pluriverse: Radical Interdependence, Autonomy, and the Making of Worlds* (Durham, NC: Duke University Press, 2018).

29. Subcomandante Marcos, "Zapatista Army of National Liberation Statement on the Kosovo War," *Left Curve* 24 (2000): 22.

30. See datasociety.net, ainow.org, newschool.edu/digital-equity-lab, data justicelab.org, and publicdatalab.org.

31. See chupadados.org.

32. See fatml.org, data-activism.net, civic.mit.edu, digitaljusticelab.ca, www.communitysolutionsva.org/files/Building_Consentful_Tech_zine .pdf, odbproject.org, and femtechnet.org/about/the-network.

33. See alliedmedia.org/amc2018/design-justice-track.

34. See Una Lee et al., "Design Justice Network's Network Principles," 2016, designjusticenetwork.org/network-principles.

35. Ito, "Resisting Reduction."

36. Paul Kingsnorth, *One No, Many Yeses: A Journey to the Heart of the Global Resistance Movement* (New York: Simon and Schuster, 2012).

4 THE FLUID BOUNDARIES OF NONCOMMUNICABLE DISEASE

Cathryn Klusmeier

Introduction

One day, around his fiftieth birthday, my father stopped talking. It wasn't a complete silence, but seemingly overnight he struggled to voice more than a few words at a time. A couple of weeks later, he quit the career he'd held for twenty years and began sleeping most of the day. Not long after, he lost his way to the bathroom. Then he lost the kitchen. He lost how to empty the dishwasher and cut the ends off asparagus and tell which door is the refrigerator. He lost how to put together puzzles and form cohesive sentences and tell you how many kids he has. If you sent him to the grocery store for toilet paper and bread, he came back with unripe mangoes and brown sugar.

After a couple years, he started seeing people who weren't there. *Murderers* he would call them. *Bad men over there.* He would stare, clench his fists, and start yelling at the crown molding or at a bush outside the window. He mistook my youngest brother for one such person, which is how my brother ended up shoved in the mud one day while they were walking in the park. After four neurologists, multiple psychiatrists, psychologists, MRIs, PET scans, a CT scan, an EEG, and multiple rounds of medications, we finally got a diagnosis. Years after he first stopped talking, a lumbar puncture came back suggesting early-onset Alzheimer's—likely developing when he was in his forties. By then, our conversations began involving discussions of towering medical bills, bankruptcy, and eviction. Without health insurance and with one parent providing fulltime care to the other parent so nobody was working, there wasn't much left.

My father did not die for nearly six years.

The first time my youngest brother told me that he too had Alzheimer's, he was fifteen years old. We were on the phone—there were thousands of miles spread between the two of us, but the connection was a good one. When he spoke those words, I was surprised at just how unsurprising that statement was. How easy the words *I have Alzheimer's* sounded in his fifteen-year-old tenor. Perhaps it was because I had heard those same words uttered by my mother—still in her forties—just a week before. *My mind is slipping*, she had said. My youngest brother was *losing things*. *Just like Dad*. He couldn't remember what was said in school; he was forgetting how to spell, how to recall basic information. *I'm serious*, he told me. My mother had said the same nearly verbatim. I had brushed it off at the time; I echoed the doctors, telling them that it was just stress, that of course a fifteen-year-old did not have Alzheimer's disease. Which is true, of course. The disease that ultimately killed my father was not diagnostically what my brother was suffering from. And yet today, just a couple years after my father's death, I'm still stuck on the words my brother uttered to me at age fifteen, the certainty in his voice. I'm less inclined to brush it off now. In fact, it is this moment, this phone call with my brother so many years ago, that is the driving force behind this essay.

Much of the literature surrounding an Alzheimer's diagnosis attends to the mechanics of the *disease* itself—the physical "unspooling" of a diseased brain via neuritic amyloid plaques, neurofibrillary tangles, and the shrinking of brain tissue (Lock 2013, 52). It's a disease that is often framed by the parameters of a singular individual, by the limits of a singular brain. However, recent literature shows that Alzheimer's, much like obesity and many other chronic, noncommunicable diseases (NCDs), is a multi-pronged disease: an interconnected and interlocking web of both individual and sociopolitical factors deeply rooted in complex cultural, economic, and political systems. In fact, current research suggests that some noncommunicable diseases are partly or even wholly *communicable*, spread through social networks, cultural conditions, and intergenerational transmission (Allen and Feigl 2017, 129). This is an essay about what it could mean to reframe noncommunicable diseases like Alzheimer's, to resist reducing them to singular individual afflictions, and instead recognize that they are complex webs of interconnected illnesses that reach far beyond the boundaries of a singular body (Ito 2017).

Spillover

"I place a fork in her right hand and guide it to the poached egg in the deep bowl. I have already cut up the toast, so that I can help her spear pieces of bread and soak up the yolk. She can't find the teacup in front of her, so I move her hand next to its handle," writes Arthur Kleinman (2009, 292). Kleinman's writing is meticulous; it is ordinary. Grounded in the everyday minutia of a dementia caregiver, it tells of loss, pain, and attention. "Caregiving is not easy," he writes. "It consumes time, energy, and financial resources. It sucks out strength and determination" (ibid., 293).

Kleinman's words are important to pay attention to here, not only because he's a physician and medical anthropologist, but because his narrative eloquently articulates some of the most enduring and pervasive tensions inherent in the life of a caregiver for someone with dementia. He locates his world within hers, within the navigation of silverware and amplification of his desperation, he too inhabits a space touched by the disease. "She is happy much of the time," he writes of his wife. "It is me, the caregiver, who, more often, is sad and despairing" (Kleinman 2009, 293) He calls his role of care a new life of "solidarity" and "enduring the unendurable" (ibid., 292–293). His exposition articulates a world where a disease of an unspooling brain—his wife's Alzheimer's—saturates and affects his life narrative in turn (Burke 2014, 30).

This process by which Kleinman's world is "saturated" by the disease of his wife is what is referred to in the literature as the *caregiver burden*—a "multidimensional response to physical, psychological, emotional, social, and financial stressors associated with the caregiving experience" (Vellone et al. 2008, 423–424). Among other health risks, "caregivers have higher rates of insomnia and depression, are at risk of serious illness, and are less likely to engage in preventative health measures" even though one half of all caregivers have at least one chronic condition (Collins and Swartz 2011, 1310). In one study, 17 percent believe that their health has deteriorated as a result of providing care, and spousal caregivers reporting high levels of strain have a 23 percent higher Framingham Stroke Risk than their non-caregiver counterparts, as well as increased all-cause mortality (as high as 63 percent in four years) (ibid.). Often, many caregivers "take on the role while healthy, but subsequently become ill" (Mittelman 2005, 634). This burden is also compounded by gender; the majority of dementia caregivers are women (Collins and Swartz 2011, 1309).

This barrage of health risks has led some researchers to note that "it might be useful to start viewing caregiving as, in itself, a health hazard" (Fonareva and Oken 2014, 725). Indeed, "one of the greatest risks for caregivers is becoming ill themselves" (Collins and Swartz 2011, 1309). The phenomenon by which a caregiver, faced with the unyielding sickness of often a spouse or a parent, *also* incurs health risks is known as the "spillover effect" (Wittenberg and Prosser 2013, 490). The person affected by spillover "may be providing care to the ill individual or may be related to the individual who is ill or both" (ibid.). The effects of spillover are broad—including somatic and psychological health, emotional health, quality of life and well-being, finances, relationship stability, and work (ibid.). Not only does the caregiver's health suffer, but the effects of this disease spill over into economics as well. Alzheimer's is among the most expensive of diseases—in 2000, the average out-of-pocket expenses for caregivers was "approximately 10 percent of the caregiver's annual income" (Collins and Swartz 2011, 1310).

While Alzheimer's is not a communicable disease in the traditional sense of disease vectors and infection, the cascading health effects of Alzheimer's upon those surrounding that individual suggest that simply framing this problem as a singular disease of a malfunctioning brain might be misleading. Using the specific language of "social networks" and "contagion effects," researchers Wittenberg and Prosser contextualize the "well-documented burden of caregiving" in the same frame as obesity and smoking, which, they note, have effects that *extend* to family members and others (Wittenberg and

Prosser 2013, 490). This language, that of extension, contagion effects, and social network systems, is important in creating a landscape of health in which caregiving and care-receiving are tightly interwoven phenomena. Overall, "research has established that health effects extend beyond one individual to include those surrounding them, including those physically present and those emotionally connected. The landscape of health decision making is altered by the conceptualization of health as a family affair" (ibid.). By invoking the "landscape of health decision making," there is some implication of context, of connection, of tightly knit social networks that together impact the health and well-being of both caregivers and receivers. Further, while "the importance of including those spillover effects has been noted by many," creating empirical estimates of spillover is challenging because the very premise of "spillover" resists compartmentalization and reductionism. It, in fact, does the opposite, creating networks of overlapping effects—social, emotional, physiological, and somatic (ibid., 498).

Perhaps one of the better demonstrations of the spillover effects of these intertwining illnesses is in the cognitive and psychological effects of caregiving. In one study detailing effects of caregiving on a spouse's cognition, researchers found that "spousal caregivers of dementia patients have on average lower levels of cognitive functioning than age, sex, and education matched controls. Caregivers performed significantly worse on measures of general cognitive functioning, speed of information processing and verbal memory" (de Vugt et al. 2006, 164; Correa et al. 2015, 371). Even further, the study found that "low performance on verbal memory was related to a decrease in caregiver competence and an increase in patient behavioral symptoms" (de Vugt et al. 2006, 164). It's important to note the ways in which cognitive impairment of the dementia patient provides a context in which the general cognitive functioning, verbal memory, and information processing of the caregivers is *in turn* impacted.

This is worth spending time on. Among a whole slew of health risks, caregivers for dementia patients are at a higher risk for *developing dementia* (Fonareva and Oken 2014, 744). In the literature of spillover, one of the more insidious problems for caregivers comes via a feedback loop whereby the effects of dementia are reflected and amplified back and forth from the dementia patient to the caregiver. "Caregiving is relational and reciprocal," writes Kleinman (2015, 240). Part of this relational and reciprocal relationship is undoubtedly the ways in which these diseases of care have spillover and "contagion-like" effects on those surrounding the disease.

Noncommunicability?

"A name that is a long-winded non-definition, and that only tells us what this group of diseases is not, is not befitting of a group of diseases that now constitutes the world's largest killer," write Allen and Feigl in a 2017 *Lancet* article discussing the utility of framing and reframing noncommunicable disease labels (Allen and Feigl 2017, 129). They begin by first describing the way diseases are traditionally divided into three groups. The first group is made up of infectious diseases like HIV, malaria, and tuberculosis. The second consists of NCDs like Alzheimer's, obesity, and cancer. The third is a category of injuries (ibid.). Calling these three divisions "outdated" and "counterproductive," they attempt to bring attention to the problematic noncommunicable disease category. They note that noncommunicable diseases "share all of the ideological and social justice issues of HIV but cause 30 times more deaths and receive 17 times less funding" (ibid). Further, even though the category begins with "non," these noncommunicable diseases "will cost the global economy US$47 trillion over the next two decades," they will "continue to push millions of people into poverty," they and will continue to be "the leading cause of death worldwide" (ibid., 129). It's not simply a matter of semantics or pedantry, they argue, because "anything that begins with 'non' must be considered a 'non-issue' or a 'non-starter" (ibid.).

The issue here is one of framing. "The disproportionately low levels of national and international attention paid to NCDs in terms of action plans, funding, and global institutions might be partly attributable to the framing of these conditions" (Allen and Feigl 2017, 129). They recognize that "calling the world's biggest killer 'non-communicable' propagates confusion, undermines efforts to spur a sense of urgency, and deflects attention from effective and system-wide interventions" (ibid.).

Beyond linguistic semantics, they also argue that the medicalized distinctions themselves need rethinking. This well-entrenched medical binary of communicable and non-communicable is a distinction that, in light of modern medical anthropological discussions, might be very tenuous. Noting that this division between communicable and noncommunicable is anything but firm, they argue that "evidence is mounting that some NCDs are partly or wholly communicable" (Allen and Feigl 2017, 129). "They can be spread through social networks, viruses such as hepatitis, cultural and economic conditions, food deserts (i.e., areas short on fresh fruit, vegetables, and other healthy foods), and intergenerational transmission (i.e., diabetes and obesity)" (ibid.). Even further, they write, "The present misnomer implies that the causes are individual rather than societal. This implication is simply not the case: NCDs have largely sociogenetic antecedents, and efforts focused on individual behaviour have little overall effect if the social and policy environments do not change in parallel" (ibid.). Despite their clear and persuasive language, Allen and Feigl (2017, 129) spend little time fleshing out what they mean when they say that "evidence is mounting that some NCDs are partly or wholly communicable." Obesity, as they mention, is one NCD that has the potential to complicate the binary of communicable and noncommunicable diseases. I contend that Alzheimer's is another such disease that resists the reductive framing of the noncommunicable label. To more fully understand how Alzheimer's fits this description, it's helpful to look at the recent research surrounding obesity as a means of seeing how an NCD might be reframed.

The move from framing obesity as a failure of individual willpower is an important shift in medical anthropology. Though there is a debate as to whether or not obesity is a disease in the first place, Annemarie Jutel specifically addresses this question, noting that where being overweight once was merely a descriptor of corpulence, it has undergone the "transition to a disease entity" (Jutel 2006, 2269). Sociologist Karen Throsby notes that often, in the contemporary rhetoric surrounding obesity, "the fat body is easily labeled as lazy, self-indulgent and lacking in discipline" (Throsby 2007, 1561). Further, "those who become fat often find themselves needing to account for their size in order to refute the suggestion of moral failure that attaches itself easily to the fat body" (ibid.). Much of the dialogue surrounding obesity reinforces this notion of an *individualized* problem and a failure of willpower. Recent obesity research, however, has begun to challenge this view of obesity, recognizing that this noncommunicable disease occurs in a sociopolitical and environmental *context*, and that there are explanations for obesity that are not reduced to the failures of individuals.

In "The Global Obesity Pandemic: Shaped by Global Drivers and Local Environments," Swinburn et al. discuss the concept of an "obesogenic environment": "The simultaneous increases in obesity in almost all countries seem to be driven mainly by changes in the global food system, which is producing more processed, affordable, and effectively marketed food than ever before" (2011, 804). Swinburn shows that the drivers of the obesity "epidemic" are inherent in the globalized food system itself. For Swinburn, obesity as disease is actually the result of people "responding *normally* to the obesogenic environments they find themselves in" (ibid.; emphasis mine). And while "support for individuals to counteract obesogenic environments will continue to be important, the priority should be for policies to reverse the obesogenic nature of these environments" (ibid, 807). Through the lens of obesogenic environments, Swinburn is able to complicate this individualized model, emphasizing that a focus on encouraging consumers to make better food choices fails to fully grasp one of the large-scale drivers of obesity: context and global markets. In a similar vein, the report "Wider Income Gaps, Wider Waistbands? An Ecological Study of Obesity and Income Equality" (Pickett et al. 2005, 670) also directs the conversation away from an individualized notion of obesity and instead focuses on the effects of poverty and income inequality on obesity. Among developed countries, "income inequality was significantly related to obesity among men and women, diabetes mortality, and average calorie intake" (ibid., 672).

Research into obesogenic environments and the impact of socioeconomic status upon obesity is important in this discussion as it relates to noncommunicable diseases, including Alzheimer's. Although it by no means shows that this is a communicable disease in the traditional sense of disease vectors, this research does complicate the boundary between communicable and noncommunicable by showing that obesity is not just an individual problem, but an interconnected and interlocking web of both individual and social factors. Obesity is not solely the result of poor individual choice, it's a multipronged disease: political, socioeconomic, and personal at the same time.

"There is an ongoing and largely unhelpful emphasis on individual healthy choices" that "hamper a shift towards more effective and equitable population-level policies such as tighter tobacco control and measures to address obesogenic environments" (Allen and Feigl 2017, 129). Allen and Feigl argue that this tendency to align noncommunicable diseases within the realm of individual bodies, such as the tendency to blame obesity on the moral failings of the individual rather than looking at larger scale factors such as obesogenic environments, obscures a more complex, nuanced understanding of the causes of disease and in turn prevents "systems-level interventions" (ibid.). Obesity, like Alzheimer's and other noncommunicable diseases, are made up of many complex and interconnected systems that contribute to the disease.

Further refuting the idea of obesity as an individualized disease, researchers have linked the spread of obesity through social networks. One study considered the extent to which weight gain in one person is associated with weight gain in friends, siblings, and spouses. Framed with the aim of looking into networks and connections as a means of studying the extent to which obesity is "spread through social ties," their findings suggest that "obesity may spread in social networks in a quantifiable and discernable pattern that depends on the nature of social ties" (Christakis and Fowler 2007, 370, 377). Their results indicate that a person's chance of becoming obese increased by 57 percent if he or she had a friend who became obese in a given interval. If one spouse became obese, the likelihood that the other spouse would become obese increased by 37 percent (ibid., 370). What's important about this study is the recognition that social networks, human interactions, and the complex web of interpersonal dynamics is now a component of obesity research much like spillover is in caregiving. As the study shows, "the spread of obesity in social networks appears to be a factor in the obesity epidemic" (ibid., 378).

These are just a few examples of the avenues in which current obesity research is finding ways to contextualize obesity and extend the etiology to social, political, and interpersonal spaces. Obesity is a complex web of relationships, and recent research suggests that trying to fit obesity into a framework that doesn't account for this nuance and complexity is problematic.

Which brings us back to Allen and Feigl's call to question the NCD category. Much of the strength of their article lies in its willingness to push against these seemingly ingrained medicalized distinctions such as communicable vs. noncommunicable and individual vs. societal. And although there is not a specific mention of Alzheimer's disease or caregiving in their article, I contend that just as obesity resists these reductionist frameworks of singular disease, so does caregiving. Alzheimer's is a disease of many interlocking problems. Like obesity, it's a disease at the nexus of political, social, economic, and interpersonal issues especially when it comes to the well-documented health risks of the caregivers, yet often the diagnosis is a highly individualized one, emphasizing the tangles and plaques of a single malfunctioning brain. Increasingly, however, the vocabulary of caregiving literature is beginning to push against the individuality. Like Allen and Feigl, I argue that the traditional framing of NCDs is problematic in the ways it decontextualizes and individualizes diseases like Alzheimer's. Further, given the literature of caregiving, I want to suggest that aspects of Alzheimer's disease resist the binary disease label that separates the communicable from the noncommunicable in the way that the diagnosis affects not only the diseased brain in question but also the health of the caregivers and the surrounding family. How does Alzheimer's stay within the noncommunicable disease boundary when the literature of caregiving uses phrases like "contagion effect" and "spillover" to describe the health effects of caregiving? (Wittenberg and Prosser 2013, 490). Indeed, Kleinman notes that the pathology of the *disease* may be limited to a single brain, but Alzheimer's clearly breaks those boundaries by making those *surrounding* the disease ill (Klienman 1980, 73).

To be clear: the argument here is not that obesity or Alzheimer's is an infectious disease or that it presents the kinds of problems that a disease like Ebola virus can engender. Insofar as I push against these communicable vs. noncommunicable labels, it is not to equate the two, as the challenges of tuberculosis are indeed very different from those of Alzheimer's. This debate is not about the utility in continuing or not continuing to use "infectious" as a way of describing particular kinds of diseases. This debate is about whether or not the reductionist language of noncommunicability, with its built-in binary, is the problem. I want to suggest that the language of infection has use, but the binary might not. Regardless, NCD is a category that is not only problematic, but might be anathema to the goals of attending to the interactional and interpersonal nature of certain diseases like Alzheimer's and obesity.

Allen and Feigl (2017, 129) note that they are not the first to call for a reevaluation of the noncommunicable and infectious disease binary, and they certainly aren't the last. Just five months after their article, two more articles were featured in *The Lancet*, one calling this reframing a "welcome" change (Rigby 2017, 653). Instead of noncommunicable, anthropologists Kozelka and Jenkins suggest renaming NCDs "interactional diseases" (Kozelka and Jenkins 2017, 655). And while the authors recognize that there is conceptual danger in conflating this notion of biological contagion with social or political contagion, Kozelka and Jenkins note that "all diseases are interactional in some sense; infectious diseases are also contracted and treated in particular social worlds" (ibid.). The "interactional nature of diseases formerly clustered as NCDs must be considered for researchers, clinicians, and policy makers to understand the complex content of interactional disease course and outcome" (ibid.). In lieu of NCDs, Allen and Feigl use the term "ecological" as a means of overcoming the binary of communicable and noncommunicable diseases (Allen and Feigl 2017, 129). By ecological they merely mean "the relation of living organisms to one another and to their physical surroundings" (ibid.). This definition, coupled with the "interactional" conception of diseases, has the potential to overcome some of the barriers inherent in the binary and, through a more ecological, relational lens, potentially reveal greater insight into how connected and complex NCDs are.

This discussion is more than just language and semantics. Given the burden of caregiving and the ways in which the disease infiltrates the social networks surrounding the person with the diseased brain, the categorizations of noncommunicability fall short. It's not just a minor issue of wording, it's a systematic failure. The framework itself is failing to account for the literature showing that to be a caregiver is to also to be at risk.

Fluid Boundaries

Anthropologist Stanley Ulijaszek uses images of water and fluidity as a guiding visualization articulating the complexities of obesity. "There are leaks and flows between bodies and the environments that surround them, as there are between disciplines that study obesity," he writes (Ulijaszek 2014, 3). "Obesity has been studied using genetic, physiological, psychological, epidemiological, cultural, environmental, political, and economic frameworks, among others" (ibid.). He notes how, increasingly, systems thinking has been applied to the study of obesity, which involves "attempts at understanding how things influence one another within a whole" (ibid.). Ulijaszek recognizes that this kind of thinking is anything but new, "since the study of ecosystems, which involves relationships among physical and biological elements within the environment, including water, goes back to the 1930s" (ibid.). Indeed, Ulijaszek writes that for obesity, conceptual frameworks that can recognize "how things influence one another within a whole" is hugely important, given that a phenomenon like obesity is "ecological, transdisciplinary, and complex" (ibid.). "Boundaries between fields of knowledge are important for their integrity (a body without boundaries would not be a body), but it is equally important that they should be leaky," he writes. "Leaky" is the word he uses to describe how one might approach such a complex issue like obesity. This language of water and fluidity, he writes, is important in that it draws attention to how scientists "should approach their pool of study: with open hands, with fluid boundaries and technologies of mind and body to try and catch the uncatchable" (ibid.).

These images of water, fluidity, permeating boundaries, and complexity can frame not only obesity but also, I argue, Alzheimer's. This notion of "spillover" has already led research in this direction—recognizing that Alzheimer's occurs in a context, and that to conceptualize the health effects of those surrounding the disease, one must consider a web of interconnected systems of social, economic, relational, and biomedical factors. And that, even then, the boundaries between those are fluid and "leaky." Margaret Lock recognizes the need to broaden the framework of Alzheimer's beyond the biomedical. She writes, "To date, relatively few researchers have paid attention to the relationship among AD incidence and poverty, social inequalities, and family histories, but this situation is beginning to change, spurred on by findings in epigenetics" (Lock 2013, 7). Medical anthropology can provide this complex framework, allowing Alzheimer's to be placed in the context of systemic poverty, social inequality, and family histories, showing how the boundaries of this disease are leaky.

The importance of contextualizing Alzheimer's patients and recognizing that their health is linked to the caregivers is clear in Collins and Swartz's "Caregiver Care." In their article, the authors lay out the health risks of the caregiver burden and talk about the potential for caregiving "interventions," which are individualized assessments where a physician simultaneously treats the caregiver and the patient with Alzheimer's (Collins and Swartz 2011, 1312). An NYU caregiver intervention program includes "individual and family counseling sessions tailored to each caregiver's specific situation, followed by weekly support group participation and ad hoc telephone counseling at the request of spouse-caregivers and other family members for the entire duration of the disease" (Mittlelman 2005, 637). Collins and Swartz note that "when patients and caregivers are treated as a dyad, outcomes for both are improved" (Collins and Swartz 2011, 1312). While it may seem like a subtle distinction, the word *dyad* holds huge importance here. When patients and caregivers are treated *together*—as a group of two people—the outcomes improve. Again, it seems subtle, but in treating this disease as a dyad, there is some recognition of the interactional nature of this complex issue.

It's not a tidy argument. The arc of this research is rooted in leaky boundaries and invocations of complexity—but that is the point. Even now, the driving force behind this essay boils down to the day my fifteen-year-old brother told me that he too had Alzheimer's, which is, of course, not strictly *true* from the biomedical lens of disease. However, if Alzheimer's can instead be seen as a web of interlocking and interpersonal illnesses, if it isn't reduced to the boundaries of a singular body, then perhaps this can account for a fifteen-year old who is worried that he too, has caught a disease he knows is "uncatchable." I draw similarities between obesity and Alzheimer's caregiving not because the research is necessarily similar, but because Alzheimer's, like obesity, requires understanding "how things influence one another within a whole" (Ulijaszek 2014, 3). They are two diseases that cannot be reduced to the problems of individualized bodies—they simply *resist that reduction*. They are interpersonal and *interconnecting* medical issues that defy linear causation and instead must be approached through the lens of fluid systems, leaky boundaries, and open hands to "catch the uncatchable" (ibid).

References

Allen, Luke N., and Andrea B. Feigl. 2017. "What's in a Name? A Call to Reframe Non-communicable Diseases." *The Lancet Global Health* 5, no. 2: e129–e130. https://doi.org/10.1016/S2214-109X(17)30001-3.

Burke, Lucy. 2014. "Oneself as Another: Intersubjectivity and Ethics in Alzheimer's Illness Narratives." *Narrative Works: Issues, Investigations, & Interventions* 4, no. 2: 28–47.

Christakis, Nicholas A., and James H. Fowler. 2007. "The Spread of Obesity in a Large Social Network over 32 Years." *New England Journal of Medicine* 357, no. 4: 370–379.

Collins, Lauren G., and Kristine Swartz. 2011. "Caregiver Care." *American Family Physician* 83, no. 11: 1309–1317.

Corrêa, M. S., K. Vedovelli, B. L. Giacobbo, C. E. B. De Souza, P. Ferrari, I. I. de Lima Argimon, J. C. Walz, F. Kapczinski, and E. Bromberg. 2015. "Psychophysiological Correlates of Cognitive Deficits in Family Caregivers of Patients with Alzheimer Disease." *Neuroscience* 286: 371–382.

de Vugt, Marjolein E., Jelle Jolles, Liesbeth van Osch, Fred Stevens, Pauline Aalten, Richel Lousberg, and Frans R. J. Verhey. 2006. "Cognitive Functioning in Spousal Caregivers of Dementia Patients: Findings from the Prospective MAASBED Study." *Age and Ageing* 35, no. 2: 160–166.

Fonareva, Irina, and Barry S. Oken. 2014. "Physiological and Functional Consequences of Caregiving for Relatives with Dementia." *International Psychogeriatrics* 26, no. 5: 725–747.

Ito, Joichi. 2017. "Resisting Reduction: A Manifesto." *Journal of Design and Science*, no. 3. https://jods.mitpress.mit.edu/pub/resisting-reduction.

Jutel, Annemarie. 2006. "The Emergence of Overweight as a Disease Entity: Measuring Up Normality." *Social Science & Medicine* 63, no. 9: 2268–2276.

Kleinman, Arthur. 1980. *Patients and Healers in the Context of Culture: An Exploration of the Borderland between Anthropology, Medicine, and Psychiatry*. Berkeley: University of California Press.

Kleinman, Arthur. 2009. "Caregiving: the Odyssey of Becoming More Human." *The Lancet* 373, no. 9660: 292–293.

Kleinman, Arthur. 2015. "Care: In Search of a Health Agenda." *The Lancet* 386, no. 9990: 240–241.

Kozelka, Ellen Elizabeth, and Janis H. Jenkins. 2017. "Renaming Noncommunicable Diseases." *The Lancet Global Health* 5, no. 7: e655.

Lock, Margaret. 2013. *The Alzheimer Conundrum: Entanglements of Dementia and Aging.* Princeton: Princeton University Press.

Mittelman, Mary. 2005. "Taking Care of the Caregivers." *Current Opinion in Psychiatry* 18, no. 6: 633–639.

Oken, Barry S., Irina Fonareva, and Helane Wahbeh. 2011. "Stress-Related Cognitive Dysfunction in Dementia Caregivers." *Journal of Geriatric Psychiatry and Neurology* 24, no. 4: 191–198.

Pickett, Kate E., Shona Kelly, Eric Brunner, Tim Lobstein, and Richard G. Wilkinson. 2005. "Wider Income Gaps, Wider Waistbands? An Ecological Study of Obesity and Income Inequality." *Journal of Epidemiology & Community Health* 59, no. 8: 670–674.

Rigby, Michael. 2017. "Renaming Non-communicable Diseases." *The Lancet Global Health* 5, no. 7: e653.

Swinburn, Boyd A., Gary Sacks, Kevin D. Hall, Klim McPherson, Diane T. Finegood, Marjory L. Moodie, and Steven L. Gortmaker. 2011. "The Global Obesity Pandemic: Shaped by Global Drivers and Local Environments." *The Lancet* 378, no. 9793: 804–814.

Throsby, Karen. 2007. "How Could You Let Yourself Get Like That?: Stories of the Origins of Obesity in Accounts of Weight Loss Surgery." *Social Science & Medicine* 65, no. 8: 1561–1571.

Ulijaszek, Stanley. 2014. "Ecologies of Water, Fluid Boundaries and Obesity Studies. *UBVO Opinion Paper Series*, 1–3. http://oxfordobesity.org/wp/wp-content/uploads/2013/10/SUopinionpaper3.pdf.

Vellone, Ercole, Giovanni Piras, Carlo Talucci, and Marlene Zichi Cohen. 2008. "Quality of Life for Caregivers of People with Alzheimer's Disease." *Journal of Advanced Nursing* 61, no. 2: 222–231.

Wittenberg, Eve, and Lisa A. Prosser. 2013. "Disutility of Illness for Caregivers and Families: A Systematic Review of the Literature. *Pharmacoeconomics* 31, no. 6: 489–500.

5 WHAT SOCIAL WORK GOT RIGHT AND WHY IT'S NEEDED FOR OUR (TECHNOLOGY) EVOLUTION

Jaclyn Sawyer

In "Resisting Reduction: A Manifesto," Joichi Ito invites us to consider a new paradigm of technology development, a collaborative model; one where we transmit our ideas, goals, dreams, and complexities to one another, person-to-person; sharing more than what is programmable into an algorithm. "It's about trying to effect change by operating at the level songs do."[1] As I read this essay I thought immediately of social work, which does this so naturally. The power of the profession lies in its unusual mix of systems thinking and heart. Social work brings dynamic thinking to some of the stickiest human problems, courage to tough conversations, and warmth to the process of making change on the individual and group level. This human-centered tenacity is the bones of a new technological era.

Who's In?

Humans—messy, complicated, emotional, rational humans—are what make our social systems so complex. In "Resisting Reduction," human complexity is described at its most basic, cellular level: "individuals themselves are systems composed of systems of systems."[2]

These already fascinatingly complex beings create families, friendships, communities, cultures, and social systems, and networks become exponentially more complex. Technology, now an embedded feature in these human systems, affects each member of a system directly or indirectly; the collateral impact on our individual lives and communities is unfathomable.

Our increasingly automated and measured society is being built with tools that are powerful enough that the misuse of them could further divide us but simultaneously offer the opportunity for deeper convergence. As we move forward, will we build technology to reflect our growing diversity? Will we build only what we currently know into our machines, or will we leave space for the unexpected, for the times we surprise ourselves or are surprised by others? And as technology evolves—daily—are we able to develop the social knowledge we require rapidly enough to make these decisions, and in the time we need to make them?

Luckily, we have experience and wisdom within our societies to lean into as we develop new technology. And there's a method for doing this that Ito describes: "'participant design'—design of systems as and by participants." This approach will steer us away from a human versus machines mentality, give us opportunities to work together and leverage diverse expertise, and let ourselves be humbled by the magnificent complexity of humankind. If we collectively pay attention to all the ways humans affect technology, and technology affects humans, we can develop a strategy for continuous learning and iteration.

But who are the participants, and how do we invite them into a collaborative model? We know the active participants: those who design technology, the end users, our peers, and networks—but what about those unseen participants not visible in our mainstream economic and social systems? I'm a social worker, and like many professionals who call themselves by this name, I work with some of the most vulnerable members of society, those who may go unseen, actively oppressed or represented in a reduced way in our data-based systems. They too are participants, and social workers can help bring their voices into system design.

I see interesting opportunities for social workers to contribute to the work ahead in human-centered technology. They are trained to work dynamically with individuals, groups, and systems. Across diverse practice areas, they bear witness to much of the nuance of the human-social experience, and their work requires them to continually adapt to changing environments, circumstances, and policies. They are trained to work with the vulnerable, the marginalized, the neglected, as well as the wealthy, the highly educated, and the well connected. They see who's in and who's missing. Social workers are continuously evaluating "goodness of fit" for a person within their environment and assessing risk and danger, all while working toward a goal of flourishing. This flexibility, paired with a healthy respect for human agency and social and racial justice, gives them unique (often overlooked) perspectives on the power and responsibility of the technological choices we make.

But we're not primed to work together. The main challenge is that we lack structures for collaboration and continual iteration. Social workers might have a valuable perspective on systems work explored in scientific disciplines, but they speak a different language, and the work is siloed. Much of social work is experiential, based on extensive field experience, and needs to be translated into something useful for technologists. Technologists have the technical ability to investigate and exploit data about social experiences and phenomena but lack the history and social context. There is a natural complement here, but we need a systematic approach to bring these groups together.

My work is focused at the intersection of social work, data, and technology, a space that gives voice to experts on the human domain in the digital realm. I lead the Data and Program Analytics department of a large nonprofit organization that provides housing opportunity for the formerly homeless. My department develops software, collects data, conducts analyses, and distributes insights to better understand the work we do and the implications in the larger field of housing and homelessness. This work is my window for observing real-world examples of a collaborative model in action. From this perspective, let me share the following thoughts on

- thinking ecologically, or how the ecological theory in social work prepares social workers to be complex systems thinkers and actors;
- predicting history, or how social welfare technology interventions can have unintended consequence; and
- working together, using an example from my own work that brought social workers, computer scientists, and analysts together to design a human-centered technology system.

Thinking Ecologically

As a discipline, social work walks an interesting line between the individual and the collective, self-determinism and environmental influence. Social work exists to foster individual well-being and healthy relationships and to promote positive interdependent exchanges between persons and their environment. Social work has a code of ethics,[3] and fundamental to this work is respect for the constellation of the profession's core values and how they are expressed within a mutual trans-action: service, social justice, human dignity, interpersonal relationships, integrity, and competence.[4]

Social work prepares its professionals with rigorous field experience and a holistic curriculum that includes theories and frameworks that question and challenge complex social systems. First-year students have a basic conceptual under-standing of feedback loops, nonlinear trajectories, and expo-sure to systems thinking. This foundation is useful for their subsequent coursework, which can take them in a variety of directions from micro to macro practice: political advocacy, behavioral economics, clinical and therapy modalities, and social enterprise, just to name a few. The practice areas and careers social workers find themselves in are numerous and diverse, and what unites them is a systems perspective.

Students are first introduced to ecological systems theory (also called the ecological model) in their first semester of graduate school in a class called Human Behavior and the Social Experience. The ecological model uses concepts from biology and the physical sciences to describe the reciprocity between individuals and their environments. The model pres-ents a means of examining how "good a fit" people are within their physical environment and how the mutual exchange of person and place affects their livelihood. In this way, there is a focus on the edges of systems, where they interact or overlap, and what happens in those spaces where one ends and another begins.

Social work training has an interesting history with systems thinking that, like the profession, evolved over time to be in alignment with the mission of the work. Systems theory was first introduced in the mid-twentieth century as a family systems model. Social workers first applied the theory, which uses biological concepts to explain the adaptation of organisms to their environments, to describe the exchange between families and their environment. Early models presented a reduced view of reciprocity where it was believed family members were influenced equally by environmental systems with equal power.[5]

Early models emphasized goodness of fit through adaptation of person and place, but did not acknowledge the influence of power dynamics, the role of the observer, and the relative power of each part of the helping system. Moving in a less reductionist direction, in the 1960s and 1970s social work systems theory assumed a more ecological approach, deconstructing the term *environment* into social determinants with varied levels of power and influence, dependent on the individual features and their desired level of connectedness to the system.[6]

In 1979, Urie Bronfenbrenner pushed ecological systems theory into practice as a model that identifies five environmental systems with and within which an individual interacts: microsystem, mesosystem, exosystem, macrosystem, and chronosystem.[7] These systems represent ecosystems and the immediate to distant influence in a person's life.

From the initial wisdom of the physical sciences, ecological theory in social work further evolved in the 1980s to incorporate pressing factors such as the influence of power and oppression, the role of the observer in recording history, and the sociocultural functions that speed up or slow down the pace at which social exchanges occur. Paramount to this theory in social work practice is an ecological perspective on race, ethnicity, and gender, and the power implicit in the transactional exchange. This is really the special sauce of social work: it does not hold a reductionist view of an individual's experience; instead, it continuously considers the features and social constructs at play in a situation, interpreted through a social justice lens.

In education and in practice, social workers are trained to see the invisible social forces that inform behavior and social organization. With this approach, a person's motivations can be as important as their actions. Trained to perceive the gestalt, they build their work sensibly and dynamically in response. In "Resisting Reduction," Ito describes an intervention approach reminiscent of social work, "Better interventions are less about solving or optimizing and more about developing a sensibility appropriate to the environment and the time. In this way they are more like music than an algorithm."[8] There is popular debate in the field of social work about whether it is more art or science. Maybe it's a synergy of art and science.[9] Practitioners bring the unseen layers of influence into the light and challenge the forces in the system not viable for the whole. They evaluate entropy in the system by identifying and deconstructing the factors that do not promote growth: intergenerational poverty, historical oppression, systemic and institutionalized racism. Social work theory aims to understand determinants to strengthen intervention, instilling an approach akin to what Ito describes as "Humility over Control." Social workers are a voice for humans in the automated age.

Predicting History

What happens when technology tools, designed for social welfare, are developed without codesign from the social service providers and the "beneficiaries" that will use them? Despite best attempts to build a peer-reviewed, evidenced-based intervention, we still witness collateral damage from technology intervention in social systems, and the people served therein.

Let's look at one of the case studies Virginia Eubanks highlights in her book *Automating Inequality*. It focuses on the coordinated entry system for housing opportunity in Los Angeles. This policy approach has gained increasing national popularity in recent years, known colloquially as "the Match .com of homeless services."[10] In a nutshell, a coordinated entry program is a systematic approach designed to connect those experiencing homelessness with the limited available housing stock in the city. Given the obvious supply and demand problem, the approach uses a survey tool, VI-SPDAT (Vulnerability Index-Service Prioritization Decision Assistance Tool) to identify the most vulnerable people in need of housing and assign a priority score for placement.[11]

From my perspective, HUD did its due diligence when assessing possible survey tools. Prior to adoption they convened expert panels[12] and assessed each survey tool on the degree to which it was valid, reliable, inclusive, person-centered, user-friendly, strengths-based, sensitive to lived experience, housing-first oriented, and transparent.[13] According to HUD, the VI-SPDAT tool is "designed to help guide case management and improve housing stability outcomes and provide an in-depth assessment that relies on the assessor's ability to interpret responses and corroborate those with evidence."[14] These criteria certainly echo some of the best practices in social work (for anyone working on the front lines of homelessness), and yet, as evidenced in *Automating Inequality*, there is a breakdown somewhere between the intention of the technology and actual impact in individuals' lives. After meeting with case managers and hearing individuals' stories, Eubanks eloquently reflects:

If homelessness is inevitable—like a disease or a natural disaster—then it is perfectly reasonable to use triage-oriented solutions that prioritize unhoused people for a chance at limited housing resources. But if homelessness is a human tragedy created by policy decisions and professional middle-class apathy, coordinated entry allows us to distance ourselves from the human impacts of our choice to not act decisively. As a system of moral valuation, coordinated entry is a machine for producing rationalization, for helping us convince ourselves that only the most deserving poor people are getting help. Those judged "too risky" are coded for criminalization. Those who fall through the cracks face prisons, institutions, or death.[15]

So how did this go astray? As Eubanks so accurately observes, homelessness is not a simple person-to-home solution. The causes (like the solutions) of homelessness are as complex as those experiencing them. Since we can't program into a machine all the reasons and conditions and actions that render people homeless, we create shortcuts: we design surveys and predictive analytics that give us a sense of objectivity and a facade of equity.

Our history follows us, even when we run in new, innovative directions, "disrupting" the status quo. This is not to criticize these technology-policy interventions; I believe we are moving in the right direction. What we still need is a deeper ecological look at complex problems, such as homelessness, with consideration for how these tools might limit, as they expand, our ability to connect people to resources. Ito brings us to the same conclusion: "Today, it is much more obvious that most of our problems—climate change, poverty, obesity and chronic disease, or modern terrorism—cannot be solved simply with more resources and greater control. That is because they are the result of complex adaptive systems that are often the result of the tools used to solve problems in the

of a data-based system. It is they who have the most to lose, and (we only hope) the most to gain from the systems we build in support.

*   *   *

I'll end here with a heartening example from my own work that brought computer scientists, social workers, and data practitioners together to codesign software that is used by homeless outreach workers. Street homeless outreach work by its nature is very data collection heavy: each interaction is recorded to store important details that will help connect an individual with services and housing and promote good communication among the team. Developing software that is easy to use, captures the correct information, and limits common data entry errors can add great value to the teams on the ground interacting with clients every day. A codesigned system informed by the end-users also helps the data analysts understand the context in which the data points exist. While this seems like the best approach to developing software and consequent analysis of the data collected therein, it can be challenging to bring these diverse practitioners together, even within a social service organization. This process can occur more naturally when using a human-centered design approach. The shared design process helps build empathy, encourages mutual understanding, and creates a common language useful for grappling with the knotty questions surrounding "how to represent people as data points." Starting with open-ended questions in a creative environment initially promotes curiosity and conversation, and ultimately it can bring creative, novel solution ideas to a shared problem.

I am continually heartened when I see social workers gaining comfort while sharing ideas about data elements, a space outside their expertise that is historically dominated by technologists (an elite few). I love seeing social workers inject person-first language into data systems—"person experiencing homelessness" rather than "homeless person"—and seeing them reinvent "unknown" options in demographic fields to describe situations that come up as they build trust with clients—"didn't want to disclose" rather than "doesn't know the answer." When they go through the exercise of imagining a database, they think about real people they know, care about, have struggled with, and they think about how this system could help them. They question why certain information is needed, challenging the ever-growing trend of "more data is better." These small suggestions reflect human-centered values and remind us that behind every data point is a real person, with joys and struggles like our own.

I am equally heartened when I see technologists, computer scientists, and data practitioners slow down their pace during the "gaining empathy phase." They begin by asking questions and gathering requirements (not different from a typical software development model), but in the context of social welfare these questions have the opportunity to expose them to more than typical end-user needs and tap into some less tangible human struggles and motivations.

The diversity of perspective going in leads to a more fully formed system with more understandable outputs. Working together from the beginning builds comfort and a relationship between these groups that makes feedback and reiteration easier (and more enjoyable). Working together creates the potential for real stories to be told from data points, and it breathes life into algorithms.

We are only at the beginning of resisting reduction in technology design. We need to construct our technological future in a way that lets us leverage human experience and compassion. For this, I encourage us to look to our colleagues in social work for guidance in keeping our technology path human-centered, paved with empathy, care, and reverence for our shared complex history of social progress.

Notes

1. Joichi Ito, "Resisting Reduction: A Manifesto," *Journal of Design and Science*, no. 3 (2017), https://jods.mitpress.mit.edu/pub/resisting-reduction.

2. Ibid.

3. National Association of Social Work, "Social Work Code of Ethics," https://www.socialworkers.org/about/ethics (accessed May 11, 2018).

4. Ibid.

5. Michael Ungar, "A Deeper, More Ecological Social Work Practice," *Social Service Review* 76, no. 3 (2002): 480–497.

6. Geoffrey G. Greif and Arthur A. Lynch, "The Eco-Systems Perspective," in *Clinical Social Work in the Eco-Systems Perspective*, ed. Carol H. Meyer (New York: Columbia University Press, 1983), 35–71.

7. Ungar, "A Deeper, More Ecological Social Work Practice," 482.

8. Ito, "Resisting Reduction."

9. Tricia B. Bent-Goodley, "The Art and Science of Social Work Revisited: Relevance for a Changing World," *Social Work* 60, no. 3 (2015): 189–190.

10. Virginia Eubanks, *Automating Inequality: How High-Tech Tools Profile, Police, and Punish the Poor* (New York: St. Martin's Press, 2015), 84.

11. OrgCode Consulting Inc. and Community Solutions, Vulnerability Index Service Prioritization Decision Assistance Tool (VI-SPDAT), American Version 2.0 for Single Adults, 2015.

12. HUD Exchange, *Assessment Tools Convening*, https://www.huduser.gov/portal/publications/reports/Assessment_tools_Convening_Rpt.html (accessed May 3, 2018).

13. HUD Exchange, *CPD-14-012: Notice on Prioritizing Persons Experiencing Chronic Homelessness and Other Vulnerable Homeless Persons in*

Permanent Supportive Housing and Recordkeeping Requirements for Documenting Chronic Homeless Status, https://www.hudexchange.info/resource/3897/notice-cpd-14-012-prioritizing-persons-experiencing-chronic-homelessness-in-psh-and-recordkeeping-requirements/ (accessed May 3, 2018).

14. OrgCode Consulting Inc., Service Prioritization Decision Assistance Tool (SPDAT), Version 4.1, http://cceh.org/wp-content/uploads/2016/07/SPDAT-v4.0-Manual.pdf (accessed March 4, 2019).

15. Eubanks, *Automating Inequality*, 123.

16. Nate Silver, *The Signal and the Noise: Why So Many Predictions Fail—but Some Don't* (New York: Penguin Press, 2012).

17. Sasha Constanza-Chock, "Design Justice, A.I., and Escape from the Matrix of Domination," *Journal of Design and Science*, no. 3 (2018), https://jods.mitpress.mit.edu/pub/costanza-chock.

6 SYSTEMS SEDUCTION

THE AESTHETICS OF DECENTRALIZATION

Gary Zhexi Zhang

> Ecology in the widest sense turns out to be the study of the interaction and survival of ideas and programs (i.e. differences, complexes of differences) in circuits.[1]

How do we deal with unimaginable complexity? Today, the prospect of ecological crisis looms over our every move, as new technologies unfurl absentmindedly into the political realm, somehow managing to disrupt a biosphere in the process. In so many areas of art and science, our situation demands that we think in terms of heterogenous systems and porous boundaries. Today, as the artist Hito Steyerl once put it, "an upload comes down as a shitstorm."[2] The 1972 publication of *The Limits to Growth*, which warned that the world system would collapse in one hundred years given "business as usual," served timely, epochal notice on our vision of exponential "progress."[3] Moreover, its use of Jay Forrester's World3 model of planetary systems dynamics prefigured our contemporary obsession with data and simulation for understanding where we are, and where we're headed. As Joichi Ito's "Resisting Reduction: A Manifesto" suggests, the once unpopular interdisciplinary science of cybernetics has returned as a paradigm through which to understand knotted social, technological, and environmental issues.[4] A cybernetic vision of open systems and regulatory feedback seems to offer a conceptual schema through which we might negotiate a more hopeful future, or at the very least, weather the shitstorm. Meanwhile, the internet has brought information networks out of the realm of military engineering and metaphysics and into the fabric of social life itself. Unpredictable networks and ecological entanglements confront us daily, from fake news to climate change, to remind us of our lack of control—a little hubris goes a long way. The challenge is to develop new strategies, polities, and intelligences that can engage in these complex systems with humility and care.

What is lost and what is found when we answer the call to think "ecosystemically"? In what follows, I want to take a step back, in order to contextualize the resurgence of the "systems approach" and its bearing on how we understand technology and society. In doing so, I consider this nebulous discourse as both an ontological enquiry and, increasingly, a design brief. In his *Theory of Moral Sentiments*, the political economist Adam Smith refers to the "spirit of system" as an "intoxicating" impulse that is "founded upon the love of humanity," yet whose trajectory can also be "[inflamed] to the madness of fanaticism." For Smith, the zealous "man of system" imagines he can "arrange different members of a great society" like pieces on a chessboard.[5] The cybernetic approach, on the other hand, invites contingency and perturbation, emphasizing dynamism and resilience in a nonlinear world. Nonetheless, the "spirit of system" is still going strong, nowhere more evidently than in the feverish discourse around blockchain, whose evangelists suggest that a new protocol will transform society for the better. Today, decentralization is the dominant paradigm through which we think about systems. To the apparent failures of central planning and the confrontations of complexity, decentralization presents itself as a sociotechnical panacea: by giving a little more agency to the parts over the whole, we could make way for emergent interactions of a truly creative kind. From asynchronous logistics to embodied intelligence, contemporary practitioners are mobilizing self-organizing behaviors to navigate, optimize, and negotiate complex ecologies. If the systems approach offers a conceptual schema for how the world works, then decentralization offers a political theory for how it *should* be organized—one that is being advocated across the ideological spectrum, from libertarian Silicon Valley capitalists like Peter Thiel to commons-oriented activists like the P2P Foundation. But what does it mean to design for the part over the whole, govern for the individual over the collective, build the platform over the society?

I call this the *aesthetics of decentralization* because it deals not with a particular set of facts, but something more like a diagram, a "spirit," and a mode of production visible across many disciplines, throughout the last century and increasingly in the present. Here I follow the philosopher Jacques Rancière's understanding of aesthetics as the "distribution of the sensible,"[6] a sensorial training through which we learn to acknowledge the world, and correspondingly, the techniques by which the world is "given" to our senses.[7] The way we see, the cultures we foster, and the technologies we build consolidate an aesthetics that defines *what we think the system is*: and in turn, our place and identity within it. These techniques demarcate what is knowable and thinkable; what is self-evident and what is left out. The development of an aesthetics can be understood as a kind of patterning, a sensorial patina that determines what is meaningful signal and what is lost to an ocean of noise.

The Seduction of Systems

The history of systems thinking is a story of desire and anxiety, as Norbert Wiener, the pioneer of cybernetics, knew well. "Like the red queen," he wrote, "we are running as fast as we can just to stay where we are."[8] Perhaps this anxiety is inevitable, as we can neither hope to control "the system" in its entirety nor absolve ourselves of our agency and let complexity do its work. Though the cybernetic approach to systems is generally associated with the dawn of information theory in the mid-twentieth century, the impulse to understand the world through a science of organization predates the invention of bits and bytes. The late nineteenth century saw a powerful tendency toward the synthesis of social theory with a materialist philosophy of nature, galvanized by techno-scientific advances and revolutionary political fervor. Following Karl Marx's "materialist conception of history," Vladimir Lenin famously proclaimed that "everything is connected to everything else."[9] Meanwhile, Alexander Bogdanov, Lenin's intellectual comrade and later his ousted political rival, was arguably the first modern systems theorist. Between 1901 and 1922, Bogdanov, a physician, philosopher, economist, science fiction writer, and revolutionary, developed a monumental work of "universal organizational science," which he called "tectology."[10] "All human activity," he wrote in 1913, "is . . . organizing or disorganizing. This means that any human activity, whether it is technical, social, cognitive or artistic, can be considered as some material of organizational experience and be explored from the organizational point of view."[11] Tectology is seldom discussed today, but readers of Wiener's cybernetics or Ludwig von Bertalanffy's general systems theory should notice deep affinities with those later sciences of organization within Bogdanov's writing.

Later, Wiener would argue: "Information is information, not matter or energy. No materialism which does not admit this can survive at the present day."[12] Though earlier monist philosophers, like Spinoza or Lucretius, had also understood nature in terms of a universal "substance," Bogdanov sought a formal theory of its regulatory dynamics, "from the point of view of the relationship among all of its parts and the relationship between the whole and its environment, i.e. all external systems."[13] Indeed, Bogdanov understood the physical realm of the natural sciences and the ethereal stuff of communication, cognition, and consciousness as part of the same living "currency," foreshadowing the expansive commodification of intangible quantities such as attention and affect by our contemporary data industries.

Bogdanov's ideas echoed a late nineteenth-century impulse toward a totalizing system of nature, combining the natural sciences with a nascent social science and moral philosophy. The term *tectology* was in fact borrowed from the German artist and naturalist Ernst Haeckel (renowned for his richly detailed illustrations of flora and fauna), who coined it to describe the "science of structures in the organic individual." For Haeckel, the organization of biological species formed part of a "world riddle," by which he understood the nature of matter and energy to be consistent with that of consciousness.[14] Meanwhile, Haeckel's contemporary in England, the biologist and polymath Herbert Spencer, developed a totalizing "synthetic philosophy" undergirded by evolutionary theory and thermodynamics. Spencer conceived of society as a "social organism"—an evolved, self-regulating system, even claiming morality to be "a species of transcendental physiology,"[15] and comparing the legal contract to the exchange of substances between the internal organs.[16] For an era captivated by the sciences of ecology and evolution, the biological meta-phor would be an enduring one, weaving human beings into the tapestry of nature, and more darkly, evincing the existing social order as an extension of "natural" law.

For Spencer, the growth of increasingly complex systems produced a "mutual dependency between parts" by which dif-ferent "organisms" could be understood by analogy. Moving fluidly between scientific inquiry and social inquiry, he mobi-lized his theories in support of a radically libertarian agenda that was at turns utilitarian, individualist, and ultimately pro-foundly conservative. A fierce critic of social reform, he viewed social welfare as enslavement to the state; societies, like species, were subject to the "survival of the fittest" (a phrase he coined), and thus develop most ideally when they are unrestrained by government. Indeed, today Spencer is perhaps best remem-bered (along with Haeckel) as one of the founding thinkers of what became social Darwinism, a discourse whose darker undertones led to eugenics. "The law of organic process," he wrote, "is the law of all progress."[17] Victorian capitalists like Andrew Carnegie took great comfort in Spencer's evolution-ism; the powerful understood their positions to be not only optimal for society, but confirmed by the natural order.[18] Spencer's immensely influential organicist "theories of every-thing" exemplified the systematic impulse of the late nine-teenth century, prefiguring the organizational sciences of the twentieth. As Wiener would later emphasize, communication and control are two sides of the same coin: the prospect of systematic knowledge through biological or statistical abstrac-tion gave "scientific" credence to grand conjectures of social structure. Mathematics turned to politics, biology to moral-ity; the systematic imaginary of biological order propagated across society and culture by the passage of translation and metaphor. Thomas Robert Malthus, for instance, whose book *An Essay on the Principle of Population* (1798) anticipated *The Limits to Growth*, concluded his grim demographic forecast with proposals for reproductive constraints on the poor.[19] (His book, in turn, had a profound influence on Darwin's theory of natural selection). Then, as now, such assured prescriptions on societal organization seldom engaged with the lives they most deeply affected or threatened the privileged status of the prescriber.

Ecology and the Rationalization of Nature

As Adam Smith observed, a utopian impulse underlies the "spirit of system." ("Whose utopia?" remains the question.) Furthermore, systems are beautiful: modern, biologically inclined theories of organization were not mechanistic, but dynamic and creative. They invoke a choreography of lively actors whose aggregate local interactions seemed to produce a universal harmony. By intimating these rhythms and cadences, systems theory promised to reveal deep structures beyond the surface of the visible world. As a boy, Wiener was himself an aspiring naturalist; he would later reflect that "diagrams of complicated structures and the problems of growth and organization . . . executed my interest fully as much as tales of adventure and discovery."[20] The enlightenment narrative of man's transcendence over nature was replaced by something arguably more sublime, a vision of humanity intricately enmeshed within the web of life. Karl Marx wrote, "What distinguishes the worst architect from the best of bees is this, that the architect raises his structure in imagination before he erects it in reality."[21] A capacity for imagination and structure invokes the two-handed nature of the systems aesthetic: one hand, held captive by wondrous complexity, and the other, raised toward abstraction, rationalization, and control. One could call the latter a technological impulse, following Heidegger's understanding of technology as a mode of "revelation."[22]

The dawn of analog electronics introduced the logical schema of electrical engineering into the ecosystemic imaginary, and, in doing so, took systems thinking from the discourses of philosophers into the hands of engineers. Circuit notation enabled the spatial representation of physical relationships through electrical schematics, lending systems theory the calculative aura of mathematical equations. In the 1950s, the pioneer of systems ecology, Howard T. Odum, developed an "energy circuit language" called *energese*. In his wide-ranging analysis of pine forests, atmospheric gas cycles, and socioeconomic systems, Odum utilized an inventory of symbols borrowed directly from electronics, while also adding a host of his own, more abstract glyphs, such as a hexagon representing "self-regulation" or dollar signs representing an economic transaction.[23] Echoing the military origins of cybernetics, these "black boxes" made ecosystemic complexes visible and operable to the minds of engineers. In turn, this diagrammatic approach would be used to form a rationalistic model of far less quantifiable systems. Odum's textbook, *Environment, Power and Society* (1971), includes an extraordinary chapter entitled "An Energetic Basis for Religion," in which he maps an ecosystemic model of moral activity. "Religion," writes Odum, "consists of programs of learned human behavior shared with other people and taught in religious institutions controlled by religious leaders."[24] In one diagram, the sun's energy flows into the realm of "good works" and "soul," which in turn is wired up to a rectangular program labeled "Natural Selection, Pearly Gates." Meanwhile, "Disordering Hell's Fire" is represented by an electrical base, connected to a constellation of symbols labeled "Realm of the Devil's Works."

As an inherently reductive methodology (Odum called his method a "detail eliminator"), the systems approach is characterized by a tension between its expansive application to ever more complex worlds that, in turn, would inevitably overflow its capture. Odum's analysis is curiously resonant with the writings of surrealist French philosopher Georges Bataille, for whom the surplus energy of society—the "accursed share"—would find its ultimate expression in the glorious excess of opulence or war.[25] In the allegory "The Solar Anus," Bataille imagines the earth as a planetary organism, sublime and abject, in the cyclical throes of erotic eruption.[26] At every moment, like entropy's "disordering fire," the ontological anxiety of chaos seeps into the sciences of control. Whether in Spencer, Odum, or Bataille, the nominally rationalistic schema was seldom more than a few steps away from theodicy. "In a very real sense," Wiener would write in *The Human Use of Human Beings* (1950), "we are shipwrecked passengers on a doomed planet."[27]

The Aesthetics of Decentralization

Like the cybernetics of Wiener and his colleagues, Odum's systems ecology invoked a world of lively matter, both living and inert. "Purposeful mechanisms," he wrote, "are self-organized into a decentralized program of ecosystem control."[28] If decentralization describes the nature of a global system without a single source of control, self-organization can be understood as the interactive local dynamics by which global order is constituted. The enduring influence of this idea proliferated across disciplines, from geology to computer science, perhaps most famously in Lynn Margulis and James Lovelock's "Gaia hypothesis," the controversial proposition that the earth is a self-regulating "organism." With the birth of cybernetics, decentralization and self-organization became not only the principles of systems theory, but the tenets of systems design and engineering. The first, and arguably most elegant, example of this was W. Ross Ashby's "homeostat." The English psychiatrist modified and connected four Royal Air Force bomb control units to produce a machine capable of responding to environmental perturbations and returning to equilibrium. For Ashby, the homeostat's "ultrastability" was analogous to the brain's capacity for learning, as well as to the evolutionary process of natural selection—adaptive behaviors within dynamic environments, whose implicate order was purposeful in appearance only.

From an administrative view, decentralization involved the automation of control. Decentered from the behavior of individuals, organization was an emergent property of the system as a whole. The idea that systems were, to some extent, essentially autonomous would be of powerful inspiration to artists, dreamers, and technocrats alike. It evinced unpredictable, responsive, and creative systems—more collaborator than instrument—producing intricate patterns of order far beyond their designers' limited prescriptions. These patterns could be found everywhere, from computational cellular automata to the distribution of human societies. Stewart Brand's countercultural "bible" of the late 1960s, the *Whole Earth Catalog*, is littered with references to chaos theory, ecological metabolisms, and "whole systems."[29] For Brand, Buckminster Fuller, and other leading futurists of the hippie generation, the beauty of self-organization affirmed the "bottom-up" transformation of society and the self against the destruction wrought by centralized governments and corporations. Self-organization gave them hope: the dissemination of technology and knowledge would engender forms of individual self-actualization they believed were necessary for a more utopian society to take shape.

In 1968, Jack Burnham, an artist and writer who was then a fellow at the Center for Advanced Visual Studies at MIT, published an essay entitled "Systems Esthetics." "We are now in transition," declared Burnham, "from an object-oriented culture to a systems-oriented culture." For Burnham, the "creation of stable, on-going relationships between organic and non-organic systems" within all "matrixes of human activity" was now the primary context for artistic and aesthetic investigation.[30] In 1970, Burnham organized *Software—Information Technology: Its Meaning for Art* at the Jewish Museum in New York. The exhibition featured leading conceptual artists of the day such as Vito Acconci and Hans Haacke, new media art pioneers such as Sonia Sheridan and Nam June Paik, and Nicholas Negroponte's Architecture Machine Group, which would later become the MIT Media Lab. Although it was, at the time, an unqualified technical and financial disaster that contributed to the dismissal of the museum's director, *Software* was a landmark experiment in which artists and technologists investigated information technology not as mere tool or entertainment, but as process and cultural paradigm. In Negroponte's contribution, *SEEK*, a group of gerbils inhabited an architectural environment made of modular blocks, which were manipulated by a robotic arm in response to the gerbils' movements. As it turned out, the gerbils were not model citizens for Negroponte's cybernetic "city," choosing instead to attack each other.[31] Ironically, Negroponte's morbid experiment exemplified the enduring influence of self-organizing, emergent principles on architects, planners, and social scientists to this day. With simple rules and responsive environments, it suggested, complexity performs itself. The "social organism" of the nineteenth century grew into the evolutionary algorithms, "soft architecture machines," and artificial societies of the information age. As the gerbils might attest, these models often stumbled over their own ambition, more reflective of the will of the designer than of intelligent design itself.

Meanwhile, over at RAND Corporation, Paul Baran was working on the schematics for a distributed communications network that would become ARPANET, the precursor to the internet. The principles of decentralized organization reified the idea that stability and control could be built into a system through its morphological, protocological, and infrastructural design. Not only were decentralized systems more resilient to perturbation, but their asynchronous logistics and self-regulating feedback could efficiently automate complex processes once relegated to burdensome (and vulnerable) centralized management. Again, Bogdanov was prescient here. In his science fiction novel *Red Star* (1908), the Soviet theorist imagines a decentralized, self-regulating economic organization known as the "Institute of Statistics." Set in a communist society on Mars, the Institute would "keep track of the flow of goods into and out of the stockpiles and monitor the productivity of all enterprises and the changes in their work forces. . . . The Institute then computes the difference between the existing and the desired situation for each vocational area and communicates the result to all places of employment. Equilibrium is soon established by a stream of volunteers."[32] Bogdanov's technocratic utopia, imagined four decades before the invention of computers, bears an uncanny resemblance to the "smart cities" of today, in which omniscient sensors and ubiquitous computing promise to solve all manner of sociotechnical challenges. In Bogdanov's city, through a non-coercive machinery of urban-scale regulation and control, "equilibrium is soon established" by the labors of a voluntary citizenry. As the historian of science Orit Halpern points out, contemporary ubiquitous computing is "imagined as necessary to supplant, and displace, the role of democratic governance."[33] Far from a socialist utopia, "futuristic" smart cities like Songdo, South Korea, are marketed to global elites as technologically enhanced Special Economic Zones, replete with financial deregulation, tax incentives, and luxury real estate.

Therein lies the contemporary dogma of decentralization. Since the early days of the Web, the design of decentralized information networks have developed in tandem with the libertarian ideal that, with technologies to ensure secure and unfettered communication between individuals, governance would organize itself. Though the early dreams of crypto-anarchy were short-lived, the dramatic and egregious centralization of power on the internet by corporations and states in the past two decades has returned the question of decentralization to the fore. The emergence of blockchain's decentralized, "trustless" networks are perhaps the most concrete iteration of this fantasy to date. Viewed energetically, "proof-of-work" implementations of blockchain automate the labor of institutional "trust" to the cryptographic infrastructure of the network, securing by algorithmic consensus and computational work, rather than the physical, political, and emotional labor involved in forming and maintaining social institutions. Similarly, smart contracts bind individuals via the insurance of executable code, deferring the social contract to an operating system.

Even if we are to ignore proof-of-work's disastrous impact on the environment, the contemporary discourse around crypto-currencies largely rests on the notion that with the right technological conditions, politics and society will follow—in this case, in the direction of individual emancipation from silos of institutional power. As journalists Michael Casey and Paul Vigna write in *The Age of Cryptocurrency*, "It speaks to the tantalizing prospect that we can take away power . . . from the banks, governments, lawyers . . . and transfer it to the periphery, to We, the People."[34] When Odum proposed his systems ecology as a "detail eliminator," he was abstracting from observable phenomena in order to bring the general picture into clearer focus. Blockchain's "trustless" utopia does the opposite, reducing the full range of human activity to game-theoretic dynamics of self-interested individuals. Where the nineteenth-century philosophers concluded that sociopolitical systems behaved in accordance with evolution and competition, these "natural laws"—and the social values they encode—are now the work of systems designers and engineers. Blockchain is libertarian to its core, built for competition over cooperation, accumulation over distribution. When political organization is conceived as a genre of game design, we need to consider the values and assumptions at play, and currently, blockchain's are powerfully skewed.

My intention here is not to dismiss the potentials of distributed ledger technologies, which clearly represent an important milestone in the development of secure, decentralized databases. Rather, it is to reject the implication that technological decentralization in our ever more informatic world is inherently aligned with a more progressive trajectory for society as a whole. Despite the cacophony of political conjecture, the story of blockchain so far is a tale of financial speculation, in which the cash rewards reaped by bankers and venture capitalists are largely the result of techno-utopian hype. *Plus ça change, plus c'est la même chose.* The prospect of decentralizing control does not absolve us of the hard work of politics, and blockchain has so far failed to transfer power to "We, the people," whatever the white papers might claim. Political economy cannot be replaced by technology alone. As Karl Marx understood over a century and a half ago, the worth we attach to technological progress is not intrinsic: it is only as valuable as the relations they produce among people. Today, technological wealth produced by society as a whole largely oils the machinery of capitalist accumulation for the few. While we have yet to witness the decentralization of control, the collective wealth produced by the decentralization of production—that is, the "sharing economy," the big data industry, and other platforms that monetize our daily social interactions—remains firmly in the service of exploitative (centralized) corporations. Whether in logistical services like Uber or social media platforms like Facebook, it is not so difficult—nor even particularly radical—to imagine decentralized, peer-to-peer services whose value is produced by and for society as a whole. Nonetheless, it would require governance, by nationalization or other means: the distributed network is not identical to the commons.

What does it mean to design decentralized systems that sit so comfortably within the regime of contemporary capitalism? If our current systems are flawed, then the technologies we build cannot be tolerant of the power structures in which we're enmeshed—attending to business as usual, albeit at an accelerated pace. "All is well since all grows better," reflected the industrialist Andrew Carnegie, happily inspired by Spencer's evolutionist thought.[35] Uncritically, the seductive power of the systems approach seems to reveal an intricate map that affirms the "nature of things" as the way they ought to be—a conservative tendency that must be resisted. As the feminist collective Laboria Cuboniks declare in its manifesto *Xenofeminist Manifesto*, "If nature is unjust, change nature!"[36]

In "A Cyborg Manifesto," the feminist technology scholar Donna Haraway describes an emancipatory figure that is "wary of holism, but needy of connection."[37] Even earlier, in 1960, the computer visionary Ted Nelson conceived of Project Xanadu, a would-be alternative to the World Wide Web that privileged "visible connections" between links. Nelson, who invented the concept of hypertext, understood from the outset that information technologies would only make us wiser if they helped us to comprehend the ways in which the complexities of our world are interconnected. These pioneering ideas remind us that rather than deferring our cognitive and political labor to increasingly automated systems, only by constantly traversing these connections can we produce a critical and reflexive understanding of how knowledge, power, and society are organized. Through this kind of systems approach, neither siding with parts nor wholes, but forever in a process of negotiation, we might realize a more emancipatory politics and its concomitant technological forms. The aesthetics of decentralization reveals a rhizomatic scene, an intuition that our routes are chaotic and ambulatory, not headlong and domineering. As Haraway wrote, over thirty years ago now, "single vision produces worse illusions than double vision or many headed monsters . . . in our present political circumstances, we could hardly hope for more potent myths for resistance and recoupling."[38] We need to imagine systems that read signals other than market signals, that answer to dreams other than Silicon Valley dreams. Contemporary transhumanists and Singularitarians should take note of Alexander Bogdanov's pioneering example one last time: the great theorist died in middle age from a botched blood transfusion, a process by which he had hoped to gain perpetual youth.

Notes

1. Gregory Bateson, *Steps to an Ecology of Mind* (New York: Ballantine Books, 1972), 491.

2. Hito Steyerl, "Too Much World: Is the Internet Dead?," *E-flux Journal #49*, November 2013, https://www.e-flux.com/journal/49/60004/too-much-world-is-the-internet-dead/.

3. Donella Meadows, Dennis Meadows, Jørgen Randers, and William Behrens III, *The Limits to Growth; A Report for the Club of Rome's Project on the Predicament of Mankind* (New York: Universe Books, 1972).

4. Joichi Ito, "Resisting Reduction: A Manifesto," *Journal of Design and Science*, no. 3 (December 2018), https://jods.mitpress.mit.edu/pub/resisting-reduction.

5. Adam Smith, *The Theory of Moral Sentiments* (Oxford: Oxford University Press, [1759] 1976), 185.

6. Jacques Rancière, *The Politics of Aesthetics: The Distribution of the Sensible*, trans. Gabriel Rockhill (London and New York: Continuum, 2004), 12.

7. The word *data* originates in the Latin "to give" or "that is given."

8. Norbert Wiener, *I Am a Mathematician* (Cambridge, MA: MIT Press, 1964), 324.

9. Quoted in Arvid Nelson, *Cold War Ecology* (New Haven: Yale University Press, 2005), xvi.

10. The word *tectology* was first coined by Ernst Haeckel to describe the "science of structures in the organic individual," though Bogdanov generalized the term. Ernst Haeckel, *The Wonders of Life: A Popular Study of Biological Philosophy*, trans. Joseph McCabe (New York and London: Harper & Brothers, 1905), 9.

11. George Gorelik, "Bogdanov's Tektology: Its Nature, Development and Influence," *Studies in Soviet Thought* 26, no. 1 (July 1983): 40.

12. Norbert Wiener, *The Human Use of Human Beings: Cybernetics and Society* (London: Free Association Books, [1950] 1989), 132.

13. Gorelik, "Bogdanov's Tektology," 40.

14. Ernst Haeckel, *Monism as Connecting Religion and Science*, trans. J. Gilchrist (London: Adam and Charles Black, 1895), 46.

15. Quoted in Walter M. Simon, "Herbert Spencer and the 'Social Organism,'" *Journal of the History of Ideas* 21, no. 2 (April–June 1960), 295.

16. Quoted in Emile Durkheim, *The Division of Labour in Society*, trans. W. D. Halls (New York: Free Press, 1997), 98.

17. Herbert Spencer, "Progress: Its Law and Cause," in *Essays: Scientific, Political and Speculative* (London: Williams and Norgate, 1891), 9.

18. Stephen Shapin, "Man with a Plan," *New Yorker*, August 13, 2017, https://www.newyorker.com/magazine/2007/08/13/man-with-a-plan.

19. Thomas Robert Malthus, *An Essay on the Principle of Population, as It Affects the Future Improvement of Society, with Remarks on the Speculations of Mr. Goodwin, M. Condorcet and Other Writers*, 1st ed. (London: J. Johnson in St. Paul's Church-Yard, 1798), https://archive.org/details/essayonprincipl00malt/page/n8 (accessed March 4, 2019).

20. Norbert Wiener, *Ex-Prodigy: My Childhood and Youth* (Cambridge, MA: MIT Press. 1953), 64.

21. Karl Marx, *Capital Volume I* (New York: The Modern Library, 1906), 198.

22. Martin Heidegger, *The Question Concerning Technology and Other Essays*, trans. William Lovitt (New York: Garland Publishing Inc., 1977), 12.

23. This is explored in depth in Irina Chernyakova, "Systems of Valuation" (MArch thesis, Massachusetts Institute of Technology, 2013).

24. Howard T. Odum, *Environment, Power and Society* (New York: Columbia University Press, [1971] 2007), 313.

25. Georges Bataille, *The Accursed Share: An Essay on General Economy*, trans. Robert Hurley (New York: Zone Books, [1949] 1988).

26. Georges Bataille, "The Solar Anus," in *Visions of Excess: Selected Writings, 1927–1939*, ed. and trans. Allan Stoekl (Minneapolis: University of Minnesota Press, 1985).

27. Wiener, *The Human Use of Human Beings*, 40.

28. Howard T. Odum, *Environment, Power and Society* (New York: Columbia University Press, [1971] 2007), 170.

29. *The Whole Earth Catalog*, Fall 1968.

30. Jack Burnham, "Systems Esthetics,'" *Artforum*, September 1968, 31.

31. Noah Wardrip-Fruin and Nick Montfort, *The New Media Reader* (Cambridge, MA: MIT Press, 2003), 253.

32. Alexander Bogdanov, *Red Star*, trans. Loren Raymond Graham and Richard Stites (Bloomington: Indiana University Press, [1908] 1984), 66.

33. Orit Halpern, *Beautiful Data* (Durham: Duke University Press, 2014), 25.

34. Michael Casey and Paul Vigna, *The Age of Cryptocurrency* (New York: Picador, 2016), 8.

35. Andrew Carnegie, *Autobiography of Andrew Carnegie* (Boston and New York: Houghton Mifflin Company, The Riverside Press Cambridge, 1920), 339.

36. Laboria Cuboniks, *The Xenofeminist Manifesto: A Politics for Alienation* (Brooklyn and New York: Verso, 2018), 82.

37. Donna Haraway, "A Cyborg Manifesto," in *Simians, Cyborgs and Women* (New York: Routledge, 1991), 151.

38. Haraway, "A Cyborg Manifesto," 154.

7 SYSTEMS JUSTICE, AI, AND THE MORAL IMAGINATION

Vafa Ghazavi

Introduction: The Inescapable Present

We live in a deeply unjust world. Across borders, life chances are hugely affected by where we are born. The median age for someone in Uganda or Niger is below sixteen years while in Germany or Japan it is around forty-six.[1] One in four children suffer stunted growth, rising to one in three in developing countries.[2] Around four billion people lack any access to the internet.[3] Within countries, social, gender, and racial disadvantages limit, often sharply, prospects for a flourishing life. Frequently such disadvantages are deeply intertwined with pervasive patterns of social and economic life. As recent research by Raj Chetty and his collaborators has shown, for instance, black boys from wealthy families in America are more likely to become poor in adulthood compared to their similarly wealthy white peers. Even when they grow up in the same neighborhood with parents at the same income level, black boys end up with lower incomes than white boys in 99 percent of the country.[4] Arbitrary limitations on "development as freedom," in Amartya Sen's famous formulation, abound.[5] And beyond inequality and oppression, though not disconnected from it, lies humanity's reckoning with climate change and what some have described as the Anthropocene. The latter concept captures the profound intensification of human impact on the environment in our age, which, according to Jedediah Purdy, "finds its most radical expression in our acknowledgment that the familiar divide between people and the natural world is no longer useful or accurate."[6]

Against this backdrop, prevailing conceptions of responsibility for realizing justice are under immense strain. In our world, causal connections to unjust harms are often highly diffuse, and the effects of discrete actions on overall outcomes are increasingly hard to discern. Even as connections between deprivations and globalized social processes have intensified, moral implications for specific agents remain unclear. The standard view of "normal justice," as the political theorist Judith Shklar called it, comes to represent the interests of some, often a relatively privileged few, while neglecting those of others.[7] Meanwhile, few alternatives to dominant market or political dogmas appear to exist that can adequately respond to these circumstances. Individual virtue seems insufficient, but institutional responses to take up the slack are not in sight either.

This is the inescapable context in which artificial intelligence is surging as a global political, economic, and cultural force. Yet this backdrop rarely makes it into the discussion of how machines ought to fit into our collective future. We cannot succumb to technological path dependencies. A defining question of our age is whether machines will exacerbate current injustices or contribute to rooting them out.

Reductionism and the Limits of Moral Mathematics

Reductionism has a firm grip on contemporary thinking about justice. At least since the influence of Descartes and Newton on scientific method in the seventeenth century, reductionism has been a powerful force on all aspects of Western thought.[8] It has also been a check on moral and political imagination. This generates a blind spot for harms that emerge from many people and organizations—most prominently corporations and states—pursuing their goals and interests within the limits of accepted rules and norms. That is, there is a blind spot in ethics for harms that emerge from interconnected, complex systems. Liberal theories of justice are generally inadequate to solve this problem. By focusing on establishing institutions to remediate and mitigate unjust outcomes without also promoting processes of social transformation needed to prevent those injustices, they narrow imagination precisely where it needs to be expanded. This is a major deficiency since—within the status quo, baseline morality of the age—either the injustices in question are rendered invisible or correlative duties to rectify them cannot be located. In other words, no one appears to be responsible for systemic harms. After all, individual components of the system are not necessarily breaching a threshold of blameworthiness, and this is what counts in much contemporary Anglo-American philosophy.

To see what is at stake, contrast the reductionist view with the vision of responsibility articulated by Simone Weil, the great mystic and intellectual, in her 1949 masterpiece on the future of France, *The Need for Roots*:

> Initiative and responsibility, to feel one is useful and even indispensable, are vital needs of the human soul . . .
>
> For this need to be satisfied it is necessary that a man should often have to take decisions in matters great or small affecting interests that are distinct from his own, but in regard to which he feels a personal concern. . . . He requires to be able to encompass in thought the entire range of activity of the social

organism to which he belongs. . . . For that, he must be made acquainted with it, be asked to interest himself in it, be brought to feel its value, its utility and, where necessary, its greatness, and be made fully aware of the part he plays in it.

Every social organism, of whatever kind it may be, which does not provide its members with these satisfactions, is diseased and must be restored to health.[9]

The liberal orthodoxy in contemporary philosophy, to say nothing of economics, does not promote this rich conception of human responsibility. It implicitly views the individual as an atomized, self-interested agent for whom moral obligations are a burden to be avoided unless and until there is a recognized violation of justice or harm.

This paradigm generates distinctive future risks. For technologists, investors, and geopolitical strategists alike, ideas of responsibility tend to be tied to notions of linear causality or narrow fiduciary roles such as the firm's responsibility to its shareholders. This approach weighs moral costs and duties as if it were an accounting ledger. It focuses on interactions between agents or the rules governing such interactions but pays little attention to system goals or paradigms. In the context of profit-driven or geopolitical competition, individual advances in AI can lead cumulatively to unanticipated harmful outcomes. Such competition is likely to intensify as the gains from AI become more central to economic and social outcomes.

The grip of this competitive dynamic threatens to co-opt, marginalize, or crowd out efforts to orient AI development toward the common good. As Stephen Cave and Seán Ó hÉigeartaigh point out, framing AI development as a "race for technological superiority" can create serious societal risks. Even if such a competitive race is not actually pursued, they argue, its narrative can generate a politics of fear or insecurity that erodes trust, limits deliberation, and dampens collaboration on an AI agenda that promotes collective benefits. Moreover, such rhetoric can itself spark a race for technological advantage.[10] Perhaps most troubling, I think, is the possibility that different actors begin to feel that such a trajectory is inevitable and adjust their moral baselines in light of that perceived reality. Once such a mindset takes hold, the prospect of pursuing an alternative paradigm diminishes. The race dynamic could become self-sustaining. What are the alternative narratives that could be promoted? Among other strategies, Cave and Ó hÉigeartaigh suggest reframing AI development as a shared priority for global good, including by emphasizing its potential to tackle large-scale challenges such as climate change and poverty. Shifting the emphasis in this way, in their view, could help counteract a race dynamic by downplaying the importance of which companies or countries make key breakthroughs, highlighting the mutual benefits of cooperation in the face of global challenges, and including the global community as stakeholders in the process of AI development.

The pressing task for our collective future with machines is therefore not simply to predict the risks, however important that might be. Nor is it to usher in a utopian Singularity. Rather, it is to imagine and then continually make and remake a world where scientific discovery and emergent technologies deepen human flourishing. This involves discerning what we value most, individually and collectively, rather than simply adjudicating ethical dilemmas or structuring society to compensate for inequalities or harms *after* the market has produced them. Even if we accept that market forces can drive productive innovation in AI, we have to ask whether the incentives embedded in the market are themselves congruent with wider social and moral purposes. Market competition left to itself cannot supply this. It is hard to resist the force of John Palfrey's argument that "it feels urgent that we examine what we care most about in humanity as we race to develop the science and technology of automation" (or, we can add, AI more broadly).[11] An ethical framework based on backward-looking blame and guilt won't be up to the great task before humanity to "design systems that participate as responsible, aware, and robust elements of even more complex systems."[12]

The dominant view of harm, even in large-scale, highly diffuse cases of systemic injustice, is one of assigning blame for discrete wrongdoing. But this model of carving up responsibility struggles when an agent's marginal contribution to a particular harm is almost unidentifiable. Moral philosophers such as Derek Parfit—a towering figure in contemporary analytic philosophy—have persuasively challenged our intuitions on this sort of problem.[13] Their arguments expose pitfalls of our "moral mathematics" and suggest that culpability of some kind is called for even when actions appear to fall below a threshold for wrongdoing. But they do little to overcome the difficulty of distinctly *systemic* harm. Even in its aggregated form, the reductionist account depends on a linear model of causality and rectification of harm, with agents connected to outcomes in predictable ways. The default becomes to blame a few exceptional wrongdoers, such as those most clearly linked to the endpoint of harm, rather than to see the wholeness of the situation.

This is unsurprising given the grip of salience and the availability heuristic on human psychology.[14] But as this type of connection diminishes or dissolves entirely through system effects that cannot be foreseen at the outset by the doers or enablers of harm, the limitations of the reductionist account become more apparent. What we need are normative reasons to creatively transform the existing incentive structure or its compliance regime, not simply mechanisms to allocate responsibility more precisely.[15] Rather than satisfying itself by closing in on a narrow set of moral duties, our theory of justice should liberate the very imaginative context that dictates the limits of harm, ethics, and virtue.

A more useful way to address this challenge is the concept of *structural injustice* developed by the political theorist Iris Marion Young before her untimely death in 2006. Structural injustice consists of social processes that "put large groups of persons under systematic threat of domination or deprivation of the means to develop and exercise their capacities, at the same time that these processes enable others to dominate or to have a wide range of opportunities for developing and exercising capacities available to them."[16] In response, Young advocated a "social connection model" of responsibility, in contrast to a legalistic liability model. According to this, "all those who contribute by their actions to structural processes with some unjust outcomes share responsibility for the injustice."[17] Young argued that liability, deriving from notions of guilt or blame for wrongdoing, is an inappropriate framework for assigning responsibility in relation to structural injustice. Liability fixates on guilt, which is unhelpful because it directs attention to some actors while absolving others, deflects attention from background conditions, and produces defensiveness, creating division where unified action is called for.

Instead, Young's model promotes *political responsibility*, which consists of an imperative to watch social institutions, monitor their effects "to make sure that they are not grossly harmful," and maintain "organized public spaces where such watching and monitoring can occur and citizens can speak publicly and support one another in their efforts to prevent suffering."[18] The meaning of politics here is "public communicative engagement with others for the sake of organizing our relationships and coordinating our actions more justly."[19]

To better grasp what this looks like in practice, consider the social transformation that has been taking place in response to the problem of modern slavery and labor exploitation in global supply chains. As Young pointed out, labor exploitation in supply chains was not widely considered a problem of the multinational firms or their consumers in the rich world until fairly recently. Rather, it was posed as a problem of unethical behavior or lax regulation/enforcement in the poorer countries where the exploitation took place. It was a moral problem for the factory manager in Bangladesh, not the CEO or consumer in New York. It took a qualitative shift in imagination supported by social movements and ordinary consumers to bring all the connected agents into the same moral frame. Indeed, the approach moved from one of interactive ethics—the ethics of each interaction in the market—to one of judging the morality of how agents are positioned within wider market and social dynamics. How we collectively view labor exploitation in global supply chains is gradually moving from a reductionist logic to one of systems, with all the political and economic implications entailed by such a move.

In thinking about the responsibility of specific agents to address structural injustices, Young argues that we ought to weigh several parameters, including an agent's power, privilege, interests (not least those of victims themselves), and ability to work collectively with others. I suggest something analogous is needed for developments in AI. To link the development of machines to human flourishing, a wide array of individual and collective agents will need to take up responsibilities to monitor system effects, call out destructive dynamics, and help model and construct alternative practices and paradigms.

It is striking that contemporary political philosophers have paid so little attention to the normative implications of complex systems. This vantage point connects agency with systemic change in a morally and psychologically compelling way with significant implications for the public sphere. Much of our thinking about justice has gone astray because it has failed to account for system effects in either the perpetuation of injustice or the realization of justice over time. Shifting the emphasis to system goals and paradigms, however, brings moral and political imagination center stage. This opens quite radical departures from prevalent thinking with regard to *who* has responsibilities for justice, *how* these responsibilities are fulfilled, and *what* the very nature of such responsibilities is. Taken together, this encourages me to propose a new way of thinking about justice, which I call *systems justice*.

Systems Justice

At the height of the global financial crisis in November 2008, on a visit to the London School of Economics, Her Majesty Queen Elizabeth II of England asked a group of assembled academics a succinct but piercing question: "Why did no one see it coming?"[20] In June of the following year, a group of leading experts met at the British Academy to try and answer the Queen's question. Their response, distilled in a letter, included this: "So in summary, Your Majesty, the failure to foresee the timing, extent and severity of the crisis and to head it off, while it had many causes, was principally a failure of the collective imagination of many bright people, both in this country and internationally, to understand the risks to the system as a whole."[21] Such failures of imagination afflict not only technocrats and economists but also the wide range of agents in complex social systems. The inability to appreciate the "system as a whole" has deep intellectual and cognitive roots. People experience systems differently to interactions with other agents, rules, and material facts. James Scott, for instance, has documented the disastrous consequences when states seduced by "high modernism" have established schemes that render complex social realities legible to control according to scientific laws.[22] More generally, humans tend to demonstrate a low aptitude for learning from complex interactions, focusing on end-state *outcomes* rather than what might have been, the *counterfactual*; we rationalize how we got to where we are as a natural and inevitable product of past events.[23] It is unsurprising, then, that we reify and naturalize existing systems.

Recognizing the ubiquity of system effects as the starting point for thinking about justice on a global scale, however, is a way out of this reductionist trap. It generates a new vantage point that unlocks the potential of each person to play their part in imagining, experimenting in, and realizing a more just world. To envision the possibilities, we can turn to the great historical struggles against slavery, racism, and patriarchy.[24] Faced with structural injustice, the response of many agents throughout history has been to seek to undermine the future durability of the system of oppression—especially its discourse and sentiments—in ways that cannot be reduced to a simple formula of obligations. Such responses do not necessarily resemble the repairing of harm to specific victims but rather look like an attack on background injustice through an expansion of collective moral imagination. In this task, relying on a precise, measurable division of responsibility within the status quo morality of formal rules and recognized institutions may even inhibit our capacity to notice unjust system-level harm. Indeed, discharging such duties can promote the illusion that we have done our fair share.

Responsibility for justice is thus tied to our *sense of self* as moral beings: there is ultimately no meaningful distinction here, I suggest, between the interests and life projects of individual agents and affirmative responsibility for justice. This suggests the paradigm should therefore be one of *coherence*, rather than dividing the costs of action. This requires citizens—including technologists, investors, and public leaders of various kinds—committed to human flourishing in a larger sense, beyond what is formally demanded of them. It is about virtue and integrity. Our theory of justice, if it is to seriously grapple with the complexity of harm in a globalized world, must make room for this constructive view of responsibility. It should encourage people to identify how they can positively apply their talents, life projects, and social roles to the transformation of unjust systems. Such a project must be reflected in the ethos of educational programs so as to enable all citizens to develop their capabilities to contribute to the process of structural transformation.

This task—one of empowerment and ennoblement—was already urgent given the challenges of global injustice and impending ecological crisis, but the advancement of AI accelerates this urgency further still. The interests driving advances in AI can either join this agenda or ignore it. But if they choose the latter, the cumulative effect will likely end up moving closer to and eventually coinciding with a resounding answer to Kwame Anthony Appiah's question: "What will future generations condemn us for?"[25]

To see what is already at stake, consider Cathy O'Neil's work on "weapons of math destruction" (or WMDs)—her descriptor for harmful mathematical models underpinning algorithms that power the data economy. As O'Neil, a data scientist and writer, acknowledges, WMDs reflect the "choices made by fallible human beings." Even when made with the best intentions, many of these models, she observes, encode "human prejudice, misunderstanding, and bias" into software systems that are now such a powerful force in everyday life.[26] What is particularly important for my argument is the way in which incentives embedded in the social system can drive these harmful outcomes through individual actions that are still within the bounds of moral acceptability. O'Neil points out the feedback and incentive for those running these models is profit: "Their systems are engineered to gobble up more data and fine-tune their analytics so that more money will pour in. Investors, of course, feast on these returns and shower WMD companies with more money."[27] Those involved in such processes invariably push for whatever marginal gains they can—the next technological advance to make that extra bit of profit—detached from the structural effects that may have a bearing on justice or human flourishing more broadly. There is a double moral blind spot here: one in terms of the immediate effects of an algorithm on identifiable people, and the other in terms of reinforcing the orientation of the system as a whole.

What is needed in response is a wider conception of responsibility that includes reenvisioning the underlying systems that allow such algorithms to proliferate without regard to their social impact. But what could ground this idea of responsibility? By acknowledging the potential of all of us in our situated social roles—as, for example, technologists, artists, intellectuals, politicians, CEOs, investors, parents, citizens—to contribute to social transformation, the project of systems justice deepens our sense of who we want to become as moral beings.

It is a provocation and invitation to do our part—humbly and with openness—in an ongoing, dynamic process of change. This approach pushes back against the premise of an atomized, self-interested agent and promotes alternative grounds for commitments to the common good.

The prevailing cost- or liability-based model suggests that some agents must be held responsible for a harm so that there is not an undue imposition on the lives of ordinary people going about their lives in morally legitimate ways. But this division is precisely what is at stake. The ends we seek must themselves be conceptualized within a given social condition in which we find ourselves. It is a moral mistake, I think, to suggest that we can partition ourselves from the world, demarcating one realm as that of personal freedom and autonomy and the other as that of social relations or the natural environment. In truth, the two are bound together. Fundamentally, systems justice starts by asking who we are and want to become as moral beings, rather than asking what costs we owe or burdens we should take on.[28] Since virtue is fortified in response to injustice, our response to injustice can itself become an aspect of the good life and human flourishing. Human interests therefore should not be viewed as fixed. As we begin to engage in system transformation, our interests and those of others can transform. After all, a flourishing life depends at least in part on our sense of something bigger than ourselves.[29]

Systems justice thus connects a bird's-eye view of justice— consistent with the "impartial spectator" deployed in Amartya Sen's conception of justice[30]—to the distinctive position of each agent in a social system. It encourages us to see the wholeness of the situation and to design institutions in light of the "admissibility of incompleteness."[31] The focus shifts from a perfectly just society derived from just institutions to a comparatively just society focused on social realizations.[32]

Systems justice, then, is neither an analytical category nor the articulation of a particular state of affairs. Rather, it is a lens through which moral agents can see the world from different vantage points and motivate their distinctive contribution to meet the moral needs of the age, including through deepening the *ethos* of justice—a concept the philosopher G. A. Cohen described as "a structure of response lodged in the motivations that inform everyday life."[33] It is a way to interpret our responsibilities given the exigencies of the multiple, nested social systems in which we live, not a formula for the design of perfect institutions. It is an ontological move that draws on complex systems as a recurring metaphor for imagining social and political life and the relationships agents have to each other over time. It is a standpoint for discerning individual system-level obligations of justice. Most importantly, it generates a practical *morality of coherence*—one that connects small-scale action and commitments with large-scale transformation—over a *compartmentalized morality*. Reading the essay by Lewis, Arista, Pechawis, and Kite (chapter 1), I am struck by how this approach mirrors Indigenous epistemologies that emphasize relationality.[34] I find this reassuring since it suggests that my theory may have particular purchase in confronting the challenges and opportunities of AI if we orient ourselves in a certain direction toward this task.

To be clear, my point is not to suggest that the direct responsibility of technologists or regulators, for example, is unimportant. Rather, it is to say that *both* direct and system responsibility matter. We should think of liability and social connection as *complementary* elements of a far-reaching vision of responsibility. Accountability can be effectively linked to systemic change.

Incentive-based measures to promote direct liability, however, must be sensitive to inadvertently shifting the logic from one of morality to one of maximizing gains, thus crowding out moral motivations.[35] Crucially, the manner and language through which perpetrators of injustice are held accountable should enrich and sharpen, rather than diminish, our sense of shared responsibility. In my approach, citizens become coauthors of justice rather than passive recipients of arrangements enacted and enforced from the top down. Analogously, holding powerful technology companies accountable is a call to wider moral agency, not a means of letting others off the hook.

It seems we do not have well-developed conceptual resources for the ambiguous space—the transitional phase—between injustice and justice. But it is in this in-between space that real-world contestation over moral ideas and their practical expression takes place. Incremental contributions are often an essential feature for these transitions.

Experimentalism and Hope

The moral implications of AI therefore require a new kind of responsibility. Agents—citizens, firms, states—must weigh their ethical responsibilities in relation to their connection to the entire system, not only discrete, linear interactions or even their simple aggregation. Systems justice subverts how we think about who has moral responsibility and how it is fulfilled. Since discourses and sentiments of the system shape it more than its mechanics, the role of artists and writers, for example, comes to the fore; small acts of moral courage become more powerful than they seem in our conventional ethical calculus.

Systems justice thus resists hasty conclusions on negligibility, the suggestion that small-scale contributions are largely irrelevant to structural outcomes. Instead, it highlights the importance of ordinary attitudes and behaviors in sustaining—or transforming—social systems. As the Plato scholar Melissa Lane suggests in her penetrating book that draws on ancient ethics to respond to contemporary challenges of environmental sustainability: "The person who embodies a new outlook becomes in virtue of that very fact a node in a new political imagination, the first step to creating a new social ethos."[36] Indeed, systems justice incorporates a moral version of the concept of the "butterfly effect" made famous by Edward Lorenz. As the feminist legal scholar Catherine MacKinnon more recently described this idea, "some extremely small simple actions, properly targeted, can come to have highly complex and large effects in certain contexts."[37] This yields a deceptively simple normative insight: the power for large-scale transformation is already latent within the system. "If no paradigm is right," Donella Meadows writes, "you can choose whatever one will help to achieve your purpose."[38]

While social scientists have sought to explain society-wide normative change through concepts such as norm, reputational, and availability *cascades* and *tipping points*,[39] these empirical phenomena have been neglected in the formulation of normative theory itself. Looking back to examples of social transformation does not yield fine-grained answers on what moral agents are obliged to do in the face of structural injustice. Rather, it suggests the wide scope moral agents have to reconceive their interests and moral identity, and the diverse ways they can direct their lives toward the realization of justice.[40] These studies reveal that the transformation of discursive and imaginative context lies within the grasp of moral agents, especially when acting in concert. They bring to light, for example, how the force of example and the generation of inertial momentum through incremental steps can reshape the social ethos and disrupt destructive path dependencies. This is most evident among an avant-garde who promote a new standard of justice,[41] but a wide array of agents, including ordinary citizens, play an indispensable role. Some of these contributions may not be salient or even visible, except perhaps retrospectively, yet this does not diminish their moral force.

What does this mean for the development of AI? At least one implication pertains to overcoming "the myth of people as socially independent," which, as Molly McCue and Kat Holmes point out, "not only limits who can participate in the system, but also who can contribute to the evolution of that system through design."[42] They pose a crucial question: "If we develop our innate ability to connect with one another as a precious resource and source of social vitality, what kind of AI could we build?"[43]

Since agents cannot predict or control the evolution of social systems ex ante, systems justice includes a normative commitment to democratic experimentalism. This principle draws on the tradition of American pragmatism found in the work of John Dewey and William James.[44] Under systems justice, institutions and citizens alike commit themselves to *discovering* harms and injustices, and innovations to reduce them over time. Systems justice thus denaturalizes current forms of the market and governance, leaving them open to revision. It harnesses people's distinctive talents and capabilities in this enterprise, promotes unorthodox alliances and multi- and "antidisciplinary" innovations,[45] and strengthens mechanisms for social learning about effective ways to reduce structural injustice and widen the development of human capabilities. Jaclyn Sawyer's essay in this volume on how social workers can shed light on history and social context so as to enrich the work of technologists illustrates what this might look like in practice.[46] Such collaboration between social workers and technologists, Sawyer argues, can help confront those heuristics that lead to a disconnect between the intention of a technology and its impact on human lives. We need much more of this kind of collaboration if AI is going to advance human flourishing.

Conclusion

Our future with machines looks set to be defined by the cumulative, often unintended consequences of many agents working to advance the frontiers of intelligence within existing rules and norms. Traditional models of responsibility are inadequate to confront this reality. The moral challenge in these circumstances is to foster citizens and design institutions that are responsive to these system-level effects. Beyond governments, agents such as individuals, civil society groups, and corporations can play a surprisingly constructive role by drawing on their distinctive talents, knowledge, and capacities. Since these agents cannot predict or control the evolution of social systems ex ante, citizens and institutions must remain open to discovering systemic injustice, innovating to root it out, and reconceiving their own interests in the process. Responsibility for systemic harm is fundamentally shared. Shirking this is not only problematic for potential victims but also corrodes the virtue and moral identity of the irresponsible agent. Ultimately, freedom for the self is connected to promoting justice for the whole since the two are permanently intertwined, part of a single moral life. Resisting reduction opens possibilities not only for extended intelligence but also for extended morality. Obligations of justice in any given moment of history—and the values we embed in the process of technological evolution—are therefore not impositions or burdens, but rather the means to become who we aspire to be as moral beings.

Notes

For discussions on various themes and stray ideas related to this essay, I am grateful to William Butler, Daniel Butt, Janina Dill, Joi Ito, Cécile Laborde, Amartya Sen, Henry Shue, Kathryn Sikkink, Roberto Mangabeira Unger, James Walsh, Jonathan Zittrain, and, especially, Jonathan Wolff.

1. "Global Health Observatory Data Repository," World Health Organization, http://apps.who.int/gho/data/view.main.POP2040ALL?lang=en.

2. "Goal 2: Zero Hunger," United Nations, https://www.un.org/sustainable development/hunger/.

3. "ICT Facts and Figures 2017," International Telecommunication Union, https://www.itu.int/en/ITU-D/Statistics/Pages/facts/default.aspx.

4. Raj Chetty, Nathaniel Hendren, Maggie R. Jones, and Sonya R. Porter, "Race and Economic Opportunity in the United States: An Intergenerational Perspective," NBER Working Paper No. 24441, March 2018.

5. Amartya Sen, *Development as Freedom* (New York: Knopf, 1999).

6. Jedediah Purdy, *After Nature: A Politics for the Anthropocene* (Cambridge, MA: Harvard University Press, 2015), 2.

7. Judith N. Shklar, *The Faces of Injustice* (New Haven and London: Yale University Press, 1990).

8. René Descartes, *A Discourse on the Method of Correctly Conducting One's Reason and Seeking Truth in the Sciences* (Oxford: Oxford University Press, [1637] 2006), 17. Descartes summarized his approach in the following way:

> I came to believe that in the place of the great number of precepts that go to make up logic, the following four would be sufficient for my purposes, provided that I took a firm and unshakeable decision never once to depart from them. The first was never to accept anything as true that I did not *incontrovertibly* know to be so; that is to say, carefully to avoid both *prejudice* and premature conclusions; and

to include nothing in my judgements other than that which presented itself to my mind so *clearly* and *distinctly*, that I would have no occasion to doubt it. The second was to divide all the difficulties under examination into as many parts as possible, and as many as were required to solve them in the best way. The third was to conduct my thoughts in a given order, beginning with the *simplest* and most easily understood objects, and gradually ascending, as it were step by step, to the knowledge of the most *complex*; and *positing* an order even on those which do not have a natural order of precedence. The last was to undertake such complete enumerations and such general surveys that I would be sure to have left nothing out.

9. Simone Weil, *The Need for Roots: Prelude to a Declaration of Duties towards Mankind* (New York: Routledge, [1949] 2002), 15.

10. Stephen Cave and Seán S. Ó hÉigeartaigh, "An AI Race for Strategic Advantage: Rhetoric and Risks," paper presented at the AAAI/ACM Conference on Artificial Intelligence, Ethics and Society, February 5, 2018, http://www.aies-conference.com/wp-content/papers/main/AIES_2018_paper_163.pdf.

11. John Palfrey, "Line-Drawing Exercises: Autonomy and Automation," *Journal of Design and Science*, no. 3 (2017), https://jods.mitpress.mit.edu/pub/issue3-palfrey.

12. Joichi Ito, "Resisting Reduction: A Manifesto," *Journal of Design and Science*, no. 3 (November 2018), https://jods.mitpress.mit.edu/pub/resisting-reduction.

13. Derek Parfit, *Reasons and Persons* (Oxford: Clarendon Press, 1984), 78–84.

14. Amos Tversky and Daniel Kahneman, "Availability: A Heuristic for Judging Frequency and Probability," *Cognitive Psychology* 5, no. 2 (September 1973): 207–232; Daniel Kahneman, *Thinking, Fast and Slow* (New York: Farrar, Straus and Giroux, 2011), 129–136.

15. See Iris Marion Young, *Responsibility for Justice* (Oxford: Oxford University Press, 2004), 375.

16. Young, *Responsibility for Justice*, 52.

17. Young, *Responsibility for Justice*, 96.

18. Young, *Responsibility for Justice*, 88.

19. Young, *Responsibility for Justice*, 179.

20. Chris Giles, "The Economic Forecasters' Failing Vision," *Financial Times*, The FT Year in Finance supplement, December 16, 2008, 5.

21. Tim Besley and Peter Hennessy, "The Global Financial Crisis—Why Didn't Anybody Notice?," *British Academy Review* 14 (November 2009): 8–10.

22. James C. Scott, *Seeing Like a State: How Certain Schemes to Improve the Human Condition Have Failed* (New Haven and London: Yale University Press, 1998).

23. Duncan J. Watts, *Everything Is Obvious: Once You Know the Answer* (New York: Crown Business, 2011).

24. On slavery, see, for example, Manisha Sinha, *The Slave's Cause: A History of Abolition* (New Haven and London: Yale University Press, 2017); Adam Hochschild, *Bury the Chains: The British Struggle to Abolish Slavery* (London: Macmillan, 2005); Neta C. Crawford, *Argument and Change in World Politics: Ethics, Decolonization, and Humanitarian Intervention* (Cambridge: Cambridge University Press, 2002). On women's suffrage (as one illustration of confronting patriarchy), see, for example, Martha Finnemore and Kathryn Sikkink, "International Norm Dynamics and Political Change," *International Organization* 52, no. 4 (Autumn 1998): 887–917; Francisco O. Ramirez, Yasemin Soysal, and Suzanne Shanahan, "The Changing Logic of Political Citizenship: Cross-National Acquisition of Women's Suffrage Rights, 1890 to 1990," *American Sociological Review* 62, no. 5 (October 1997): 735–745. One of the lessons from these historical cases is that the more granular perspective we take, the more it becomes apparent that ordinary citizens have been crucial to transformational change even if less valorized compared to a few exceptional leaders. For example, it might easily be forgotten that around the 1790s, 300,000 English people participated in a sugar boycott to abolish slavery (see Hochschild, *Bury the Chains*, 192–196).

25. Kwame Anthony Appiah, "What Will Future Generations Condemn Us For?," *Washington Post*, September 26, 2010, B01; Kwame Anthony Appiah, *The Honor Code: How Moral Revolutions Happen* (New York and London: W. W. Norton and Company, 2010).

26. Cathy O'Neil, *Weapons of Math Destruction: How Big Data Increases Inequality and Threatens Democracy* (London: Allen Lane, 2016), 3.

27. O'Neil, *Weapons of Math Destruction*, 13.

28. On the application of this to the public realm, consider the argument in Benjamin R. Barber, *Strong Democracy: Participatory Politics for a New Age* (Berkeley, Los Angeles, and London: University of California Press, [1984] 2003). For instance, on reductionist conceptions of freedom that pose it in opposition to power, Barber remarks: "Rendering freedom and power in physical terms not only misconstrues them, it produces a conception of political liberty as entirely passive. Freedom is associated with the unperturbedness of the inertial body, with the motionless of the inertial frame itself. It stands in stark opposition to the idea of politics as activity, motion, will, choice, self-determination, and self-realization. . . . The modern liberal appears to regard it [tranquility] as a republican ideal: man at rest, inactive, nonparticipating, isolated, uninterfered with, privatized, and thus free" (36).

29. For one argument that offers this view from the perspective of psychology, see Martin E. P. Seligman, *Flourish: A Visionary New Understanding of Happiness and Well-Being* (New York: Free Press, 2011).

30. Sen himself borrows this from Adam Smith. See Amartya Sen, *The Idea of Justice* (Cambridge, MA: Harvard University Press, 2009), 124–152.

31. Sen, *The Idea of Justice*, 131.

32. Sen, *The Idea of Justice*, 134.

33. G. A. Cohen, *Rescuing Justice and Equality* (Cambridge, MA: Harvard University Press, 2008), 123.

34. Jason Edward Lewis, Noelani Arista, Archer Pechawis, and Suzanne Kite, "Making Kin with the Machines," *Journal of Design and Science*, no. 3 (2018), https://doi.org/10.21428/bfafd97b.

35. See Samuel Bowles, *The Moral Economy: Why Good Incentives Are No Substitute for Good Citizens* (New Haven: Yale University Press, 2016).

36. Melissa Lane, *Eco-Republic: What the Ancients Can Teach Us about Ethics, Virtue, and Sustainable Living* (Princeton and Oxford: Princeton University Press, 2012), 64.

37. Catherine A. MacKinnon, *Butterfly Politics* (Cambridge, MA, and London: Harvard University Press, 2017), 1. MacKinnon draws on this concept to name her book, which compiles interventions made over

forty years. She uses the term "butterfly politics" as "an organizing metaphor and central conceit" for the volume.

38. Donella H. Meadows, *Thinking in Systems: A Primer*, ed. Diana Wright (White River Junction, VT: Chelsea Green Publishing, 2008), 164.

39. Seminal examples include Timur Kuran, *Private Truths, Public Lies: The Social Consequences of Preference Falsification* (Cambridge, MA: Harvard University Press, 1997); Cass R. Sunstein, "Social Roles and Social Norms," *Columbia Law Review* 96, no. 4 (May 1996): 903–968; Timur Kuran and Cass R. Sunstein, "Availability Cascades and Risk Regulation," *Stanford Law Review* 51, no. 4 (April 1999): 683–768.

40. To illustrate this, consider how James Baldwin sought to resolve the dilemma he faced when deciding how to play his part in responding to racial injustice in the United States in the mid-twentieth century. After listing other avenues of black resistance and activism and explaining how he did not feel personal compatibility with them, Baldwin writes: "'This was sometimes hard on my morale, but I had to accept, as time wore on, that part of my responsibility—as a witness—was to move as largely and as freely as possible, to write the story, and to get it out." See James Baldwin and Raoul Peck, *I Am Not Your Negro* (New York: Vintage Books, 2017), 31.

41. Lea Ypi, *Global Justice and Avant-Garde Political Agency* (Oxford: Oxford University Press, 2012).

42. Molly McCue and Kat Holmes, "Myth and the Making of AI," *Journal of Design and Science*, no. 3 (2018), https://jods.mitpress.mit.edu /pub/holmes-mccue.

43. McCue and Holmes, "Myth and the Making of AI."

44. See, for example, John Dewey, *The Public and Its Problems: An Essay in Political Inquiry* (University Park: The Pennsylvania State University Press, [1927] 2012); William James, *The Will to Believe and Other Essays in Popular Philosophy* (Cambridge, MA, and London: Harvard University Press, [1897] 1979). For more contemporary thinking related to this approach, see Christopher K. Ansell, *Pragmatist Democracy: Evolutionary Learning as Public Philosophy* (Oxford and New York: Oxford University Press, 2011); Michael C. Dorf and Charles F. Sabel, "A Constitution of

Democratic Experimentalism," *Columbia Law Review* 98, no. 2 (March 1998): 267–473; Charles F. Sabel and Jonathan Zeitlin, "Experimentalist Governance," in *The Oxford Handbook of Governance*, ed. David Levi-Faur (Oxford: Oxford University Press, 2012), 169–184.

45. On the idea of an antidisciplinary research program, see Joichi Ito, "Design and Science," *Journal of Design and Science*, no. 1 (2017), https://jods.mitpress.mit.edu/pub/designandscience.

46. Jaclyn Sawyer, "What Social Work Got Right and Why It's Needed for Our (Technology) Evolution," *Journal of Design and Science*, no. 3 (2018), https://jods.mitpress.mit.edu/pub/sawyer.

APPENDIX

RESISTING REDUCTION: A MANIFESTO

Joichi Ito

Nature's ecosystem provides us with an elegant example of a complex adaptive system where myriad "currencies" interact and respond to feedback systems that enable both flourishing and regulation. This collaborative model—rather than a model of exponential financial growth or the Singularity, which promises the transcendence of our current human condition through advances in technology—should provide the paradigm for our approach to artificial intelligence. More than sixty years ago, MIT mathematician and philosopher Norbert Wiener warned us that "when human atoms are knit into an organization in which they are used, not in their full right as responsible human beings, but as cogs and levers and rods, it matters little that their raw material is flesh and blood."[1] We should heed Wiener's warning.

The Whip That Lashes Us

The idea that we exist for the sake of progress, and that progress requires unconstrained and exponential growth, is the whip that lashes us. Modern companies are the natural product of this paradigm in a free market capitalist system. Wiener called corporations "machines of flesh and blood" and automation "machines of metal." These machines of bits—the new species of Silicon Valley megacompanies—are developed and run in great part by people who believe in a new religion, Singularity. This new religion is not a fundamental change in the paradigm, but rather the natural evolution of the worship of exponential growth applied to modern computation and science. The asymptote^2 of the exponential growth of computational power is artificial intelligence.

The notion of Singularity—that AI will supercede humans with its exponential growth, and that everything we have done until now and are currently doing is insignificant in comparison—is a religion created by people who have the experience of using computation to solve problems heretofore considered impossibly complex for machines. They have found a perfect partner in digital computation—a seemingly knowable, controllable indeterminate system of thinking and creating that is rapidly increasing in its ability to harness and process complexity, bestowing wealth and power on those who have mastered it. In Silicon Valley, the combination of groupthink and the financial success of this cult of technology has created a positive feedback system that has very little capacity for regulating through negative feedback. While they would resist having their beliefs compared to a religion and would argue that their ideas are science- and evidence-based, those who embrace Singularity engage in quite a bit of arm waving and make leaps of faith based more on trajectories than on ground truths to achieve their ultimate vision.

Singularitarians believe that the world is "knowable" and computationally simulatable, and that computers will be able to process the messiness of the real world just as they have every other problem that everyone said couldn't be solved by computers. To them, this wonderful tool, the computer, has worked so well for everything so far that it must continue to work for every challenge we throw at it, until we have transcended known limitations and ultimately achieve some sort of reality escape velocity. Artificial intelligence is already displacing humans in driving cars, diagnosing cancers, and researching court documents. The idea is that AI will continue this progress and eventually merge with human brains and become an all-seeing, all-powerful superintelligence. For true believers, computers will augment and extend our thoughts into a kind of "amortality." (Part of Singularity is a fight for such amortality, the idea that while one may still die, one's death will not be not the result of the grim reaper of aging.)

But if corporations are a precursor to our transcendance, the Singularitarian view that with more computing and biohacking we will somehow solve all of the world's problems, or that the Singularity will solve us, seems hopelessly naive. As we dream of the day when we have enhanced brains and amortality and can think big, long thoughts, corporations already have a kind of amortality. They persist as long as they are solvent and they are more than a sum of their parts—arguably an amortal superintelligence.

More computation does not makes us more "intelligent," only more computationally powerful.

For Singularity to have a positive outcome requires a belief that, given enough power, the system would somehow figure out how to regulate itself. The final outcome would be so complex that while we humans couldn't understand it now, it would understand and solve itself. Some believe in something that looks a bit like the former Soviet Union's master planning but with full information and unlimited power. Others have a more sophisticated view of a distributed system. But at some level, all Singularitarians believe that with enough power and control, the world is "tamable." Not all who believe in Singularity worship it as a positive transcendence bringing immortality and abundance, but they do believe that a judgment day is coming when all curves go vertical.

Whether you are on an S-curve or a bell curve, the beginning of the slope looks a lot like an exponential curve. To systems dynamics people, an exponential curve shows self-reinforcement—that is, a positive feedback curve without limits. Maybe this is what excites Singularitarians and scares systems people. Most people outside the Singularity bubble believe in S-curves: nature adapts and self-regulates, and, for example, when a pandemic has run its course, growth slows and things adapt. They may not be in the same state, and a phase change could occur, but the notion of Singularity—especially as some sort of savior or judgment day that will allow us to transcend the messy, mortal suffering of our human existence—is fundamentally a flawed one.

This sort of reductionist thinking isn't new. When B. F. Skinner discovered the principle of reinforcement and was able to describe it, we designed education around his theories. Learning scientists know now that behaviorist approaches work for only a narrow range of learning, but many schools continue to rely on drill and practice. Take, for another example, the eugenics movement, which greatly and incorrectly oversimplified the role of genetics in society. This movement helped fuel the Nazi genocide by providing a reductionist scientific view that we could "fix humanity" by manually pushing natural selection. The echoes of the horrors of eugenics exist today, making almost any research trying to link genetics with things like intelligence taboo.

We should learn from our history of applying over-reductionist science to society and try to, as Wiener says, "cease to kiss the whip that lashes us." While it is one of the key drivers of science—to elegantly explain the complex and reduce confusion to understanding—we must also remember what Albert Einstein said: "Everything should be made as simple as possible, but no simpler."[3] We need to embrace the unknowability—the irreducibility—of the real world that artists, biologists, and those who work in the messy world of liberal arts and humanities are familiar with.

Introduction: The Cancer of Currency

As the sun beats down on Earth, photosynthesis converts water, carbon dioxide, and the sun's energy into oxygen and glucose. Photosynthesis is one of the many chemical and biological processes that transform one form of matter and energy into another. These molecules then get metabolized by other biological and chemical processes into yet other molecules. Scientists often call these molecules "currencies" because they represent a form of power that is transferred between cells or processes to mutual benefit—"traded," in effect. The biggest difference between these and financial currencies is that there is no "master currency" or "currency exchange." Rather, each currency can be used by only certain processes, and the "market" of these currencies drives the dynamics that are "life."

As certain currencies became abundant as an output of a successful process or organism, other organisms evolved to take that output and convert it into something else. Over billions of years, this is how Earth's ecosystem has evolved, creating vast systems of metabolic pathways and forming highly complex self-regulating systems that, for example, stabilize our body temperatures or the temperature of the earth, despite continuous fluctuations and changes among the individual elements at every scale—from micro to macro. The output of one process becomes the input of another. Ultimately, everything interconnects.

We live in a civilization in which the primary currencies are money and power—where more often than not, the goal is to accumulate both at the expense of society at large. This is a very simple and fragile system compared to Earth's ecosystems, where myriads of "currencies" are exchanged among processes to create hugely complex systems of inputs and outputs with feedback systems that adapt and regulate stocks, flows, and connections.

Unfortunately, our current human civilization does not have the built-in resilience of our environment, and the paradigms that set our goals and drive the evolution of society today have set us on a dangerous course that Wiener warned us about decades ago. The paradigm of a single master currency has driven many corporations and institutions to lose sight of their original missions. Values and complexity are focused increasingly on prioritizing exponential financial growth, led by for-profit corporate entities that have gained autonomy, rights, power, and nearly unregulated societal influence. The behavior of these entities is akin to cancer. Healthy cells regulate their growth and respond to their surroundings, even eliminating themselves if they wander into an organ where they don't belong. Cancerous cells, on the other hand, optimize for unconstrained growth and spread with disregard to their function or context.

We Are All Participants

The Cold War era, when Wiener was writing *The Human Use of Human Beings*, was defined by the rapid expansion of capitalism and consumerism, the beginning of the space race, and the coming of age of computation. It was a time when it was easier to believe that systems could be controlled from the outside, and that many of the world's problems would be solved through science and engineering.

The cybernetics that Wiener primarily described during that period were concerned with feedback systems that could be controlled or regulated from an objective perspective. This so-called first-order cybernetics assumed that the scientist as the observer could understand what was going on, thereby enabling the engineer to design systems based on observation or insight from the scientist.

Today, it is much more obvious that most of our problems—climate change, poverty, obesity and chronic disease, and modern terrorism—cannot be solved with simply more resources and greater control. That is because they are the result of complex adaptive systems that are often the result of the tools used to solve problems in the past, such as endlessly increasing productivity and attempts to control things. This is where second-order cybernetics comes into play—the cybernetics of self-adaptive complex systems, where the observer is also part of the system itself. As Kevin Slavin says in "Design as Participation," "You're not stuck in traffic you are traffic."[4]

In order to effectively respond to the significant scientific challenges of our times, I believe we must view the world as many interconnected, complex, self-adaptive systems across scales and dimensions that are unknowable and largely inseparable from the observer and the designer. In other words, we are participants in multiple evolutionary systems with different fitness landscapes[5] at different scales, from our microbes to our individual identities to society and our species. Individuals themselves are systems composed of systems of systems, such as the cells in our bodies that behave more like system-level designers than we do.

While Wiener does discuss biological evolution and the evolution of language, he doesn't explore the idea of harnessing evolutionary dynamics for science. Reproduction and survival have driven the biological evolution of individual species (genetic evolution), instilling in us goals and yearnings to procreate and grow. That system continually evolves to regulate growth, increase diversity and complexity, and enhance its own resilience, adaptability, and sustainability.[6] As designers with growing awareness of these broader systems, we have goals and methodologies defined by the evolutionary and environmental inputs from our biological and societal contexts. But machines with emergent intelligence have discernibly different goals and methodologies. As we introduce machines into the system, they will augment not only individual humans but also—and more importantly—complex systems as a whole.

Here is where the problematic formulation of "artificial intelligence" becomes evident, as it suggests forms, goals, and methods that stand outside interaction with other complex adaptive systems. Instead of thinking about machine intelligence in terms of humans versus machines, we should consider the system that integrates humans and machines—not artificial intelligence, but extended intelligence. Instead of trying to control or design or even understand systems, it is more important to design systems that participate as responsible, aware, and robust elements of even more complex systems. And we must question and adapt our own purpose and sensibilities as designers and components of the system with a much more humble approach: humility over control.

We could call it "participant design"—design of systems as and by participants—that is more akin to the increase of a flourishing function, where flourishing is a measure of vigor and health rather than scale or power. We can measure the ability of systems to adapt creatively, in addition to their resilience and their ability to use resources in interesting ways.

Better interventions are less about solving or optimizing and more about developing a sensibility appropriate to the environment and the time. In this way, they are more like music and less like an algorithm. Music is about a sensibility or "taste," with many elements coming together into a kind of emergent order. Instrumentation can nudge or cause the system to adapt or move in an unpredictable and unprogrammed manner, while still making sense and holding together. Using music itself as an intervention is not a new idea; in 1707, Andrew Fletcher, a Scottish writer and politician, said, "Let me make the songs of a nation, and I care not who makes its laws."

If writing songs instead of laws feels frivolous, remember that songs typically last longer than laws, have played key roles in various hard and soft revolutions, and end up being transmitted person-to-person along with the values they carry. It's not about music or code. It's about trying to effect change by operating at the level songs do. Donella Meadows, among others, articulates this in her book *Thinking in Systems*.

Meadows, in her chapter "Leverage Points: Places to Intervene in a System," describes how we can intervene in a complex, self-adaptive system. For her, interventions that involve changing parameters or even changing the rules are not nearly as powerful or as fundamental as changes in a system's goals and paradigms.[7]

When Wiener discussed our worship of progress, he said: "Those who uphold the idea of progress as an ethical principle regard this unlimited and quasi-spontaneous process of change as a Good Thing, and as the basis on which they

PLACES TO INTERVENE IN A SYSTEM
(in increasing order of effectiveness)

12. Constants, parameters, numbers (such as subsidies, taxes, standards).

11. The sizes of buffers and other stabilizing stocks, relative to their flows.

10. The structure of material stocks and flows (such as transport networks, population age structures).

9. The lengths of delays, relative to the rate of system change.

8. The strength of negative feedback loops, relative to the impacts they are trying to correct against.

7. The gain around driving positive feedback loops.

6. The structure of information flows (who does and does not have access to information).

5. The rules of the system (such as incentives, punishments, constraints).

4. The power to add, change, evolve, or self-organize system structure.

3. The goals of the system.

2. The mindset or paradigm out of which the system — its goals, structure, rules, delays, parameters — arises.

1. The power to transcend paradigms.

guarantee to future generations a Heaven on Earth. It is possible to believe in progress as a fact without believing in progress as an ethical principle; but in the catechism of many Americans, the one goes with the other."[8] Instead of discussing the lack of "sustainability" as something to be "solved" in the context of a world where bigger is still better and more than enough is not too much, perhaps we should examine the values and the currencies of the fitness functions[9] and consider whether they are suitable and appropriate for the systems in which we participate.

Conclusion: A Culture of Flourishing

Developing a sensibility and a culture of flourishing—a term that has taken on special significance since the field of virtue ethics arose from Elizabeth Anscombe's 1958 essay[10]—and embracing a diverse array of measures of "success" depend less on the accumulation of power and resources and more on diversity and the richness of experience. This is the paradigm shift we need. This will provide us with a wealth of technological and cultural patterns to draw from to create a highly adaptable society. This diversity also allows the elements of the system to feed each other without the exploitation and extraction ethos created by a monoculture with a single currency. It is likely that this new culture will spread as spirituality, fashion, music, or other forms of art.

As a native of Japan, I am heartened by a group of junior high school students I spoke to there recently who, when I challenged them about what they thought we should do about the environment, asked questions about the meaning of happiness and the role of humans in nature. I am likewise heartened to see many of my students at the MIT Media Lab and in the Principles of Awareness class that I co-teach with the Venerable Tenzin Priyadarshi using a variety of metrics (currencies) to measure their success and meaning and grappling directly with the complexity of finding one's place in our complex world.

I'm also heartened by organizations such as the IEEE, which is initiating design guidelines for the development of artificial intelligence around human well-being instead of around economic impact. The work by Peter Seligmann, Christopher Filardi, and Margarita Mora from Conservation International is creative and exciting because it approaches conservation by supporting the flourishing of Indigenous people—not undermining it. Another heartening example is that of the Shinto priests at the Ise Grand Shrine, who have been planting and rebuilding the shrine every twenty years for the last 1,300 years in celebration of the renewal and the cyclical quality of nature.

In the 1960s and 1970s, the hippie movement tried to pull together a Whole Earth movement, but then the world swung back toward the consumer and consumption culture of today. I hope and believe that a new awakening will happen and that a new sensibility will cause a nonlinear change in our behavior through a cultural transformation. While we can and should continue to work at every layer of the system to create a more resilient world, I believe the cultural layer is the layer with the most potential for a fundamental correction away from the self-destructive path that we are currently on. I think that it will yet again be about the music and the arts of young people reflecting and amplifying a new sensibility: a turn away from greed to a world where "more than enough is too much," and in which we can flourish in harmony with nature rather than through the control of it.

Notes

1. Norbert Wiener, *The Human Use of Human Beings: Cybernetics and Society* (Boston: Houghton Mifflin, 1954), 185.

2. An asymptote is a line that continually approaches a given curve but does not meet it at any finite distance. In Singularity, this is the vertical line that occurs when the exponential growth curve approaches a vertical line. There are more arguments about where this asymptote is among believers than about whether it is actually coming.

3. This is a common paraphrase. What Einstein actually said was, "It can scarcely be denied that the supreme goal of all theory is to make the irreducible basic elements as simple and as few as possible without having to surrender the adequate representation of a single datum of experience." See Andrew Robinson, "Did Einstein Really Say That?," *Nature*, April 30, 2018, https://www.nature.com/articles/d41586-018-05004-4.

4. Western philosophy and science is dualistic as opposed to the more "Eastern" nondualistic approach. A whole essay could be written about this, but the idea of a subject/object or a designer/designee is partially linked to the notion of self in Western philosophy and religion. See Kevin Slavin, "Design as Participation," *Journal of Design and Science*, no. 1 (February 2016), https://jods.mitpress.mit.edu/pub/design-as-participation. The phrase originated in a drive-time conversation about contemporary design with Joi Ito.

5. Fitness landscapes arise when you assign a fitness value for every genotype. The genotypes are arranged in a high-dimensional sequence space. The fitness landscape is a function of that sequence space. In evolutionary dynamics, a biological population moves over a fitness landscape driven by mutation, selection and random drift. See M. A. Nowak, *Evolutionary Dynamics: Exploring the Equations of Life* (Cambridge, MA: Harvard University Press, 2006).

6. Nowak, *Evolutionary Dynamics*.

7. Donella H. Meadows, "Leverage Points: Places to Intervene in a System," in *Thinking in Systems: A Primer*, ed. Diana Wright (White River Junction, VT: Chelsea Green Publishing, 2008), 145–165.

8. Wiener, *The Human Use of Human Beings*, 42.

9. A fitness function is a function that is used to summarize, as a measure of merit, how close a solution is to a particular aim. It is used to describe and design evolutionary systems.

10. G. E. M. Anscombe, "Modern Moral Philosophy," *Philosophy* 33, no. 124 (January 1958): 1–19. This article is usually taken as the start of modern virtue ethics that has revived Aristotelian ethics. Virtue ethics asks what makes a good life: how can we blossom, grow, flourish?

CONTRIBUTORS

Noelani Arista is Associate Professor of Hawaiian and U.S. History at University of Hawai'i-Mānoa. Her research and writing centers on translation and indigenizing research practice in Hawaiian language textual and digital archives. Her work focuses on governance and the practice of *mo'olelo Hawai'i* (Hawaiian history). She is the author of *The Kingdom and the Republic: Sovereign Hawai'i and the Early United States* (2019).

Sasha Costanza-Chock is a scholar, activist, and media-maker, and currently Associate Professor of Civic Media at MIT. They are a Faculty Associate at the Berkman-Klein Center for Internet & Society at Harvard University, and creator of the MIT Codesign Studio (codesign.mit.edu). Their work focuses on social movements, transformative media organizing, and design justice. Costanza-Chock's first book, *Out of the Shadows, Into the Streets: Transmedia Organizing and the Immigrant Rights Movement* was published by the MIT Press in 2014.

Kate Darling is a research specialist at the MIT Media Lab and a leading expert in robot ethics and policy. Her research interests include social robotics, human–robot interaction, and intellectual property. Darling's work has been featured in *Vogue*, the *New Yorker*, *The Guardian*, BBC, NPR, PBS, the *Boston Globe*, *Forbes*, CBC, *WIRED*, *Boston Magazine*, *The Atlantic*, *Slate*, *Die Zeit*, the *Japan Times*, *Robohub*, *IEEE*, and more. She is the author of *The New Breed: What Our History with Animals Reveals About Our Future with Robots* (2021).

Vafa Ghazavi is a doctoral candidate and John Monash Scholar at Balliol College, University of Oxford. His research focuses on moral and political theory.

Kite aka Suzanne Kite is an Oglala Lakota performance artist, visual artist, and composer raised in Southern California. Her research is concerned with contemporary Lakota mythologies and epistemologies and investigates contemporary Lakota storytelling through research-creation, computational media, and performance practice. Recently, Kite has been developing body and braid interfaces for movement performances. She is currently a PhD student at Concordia University.

Cathryn Klusmeier is a writer living and working in Sitka, Alaska. She has a master's degree in medical anthropology from the University of Oxford and is currently pursuing her MFA in creative writing at the University of Iowa where she is an Iowa Arts Fellow. With a background in environmental studies and art, she often focuses on the links between landscapes, climate, and medicine.

Jason Edward Lewis is University Research Chair in Computational Media and the Indigenous Future Imaginary and Professor of Computation Arts at Concordia University, Montreal. He co-directs Aboriginal Territories in Cyberspace, the Skins Workshops on Aboriginal Storytelling and Video Game Design, and the Initiative for Indigenous Futures. Born and raised in Northern California, Lewis is Cherokee, Hawaiian, and Samoan.

Archer Pechawis is a performance artist, new media artist, filmmaker, writer, curator, and educator. He has been a practicing artist since 1984 with particular interest in the intersection of Plains Cree culture and digital technology, merging "traditional" objects such as hand drums with digital video and audio sampling. Of Cree and European ancestry, he is a member of Mistawasis First Nation, Saskatchewan.

Jaclyn Sawyer is a citizen scholar, social worker, and data practitioner, passionate about using small data in thoughtful ways. She currently leads the data team of a nonprofit that provides homeless outreach and housing opportunity. Previously she has worked on projects that expand data analytics capacity in prison reform, health care, and education in the United States, Nigeria, Nicaragua, and Brazil. She loves design thinking and uses it in everything she does.

Gary Zhexi Zhang is an artist and writer, working in film, installation, and software. Recent exhibitions include *Cross-feed* at Glasgow International 2018, vdrome.org (online), and *All Channels Open* at Wysing Arts Centre. Zhang is a regular contributor to *Frieze* and *Elephant* magazines and has also published in *Foam*, *Fireflies*, and *King's Review*. He is a graduate of the MIT Program in Art, Culture and Technology.

Snoweria Zhang is a designer, artist, and mathematician. Currently, her work focuses on the interstitial space between the built environment and the digital fabric of cities. Most recently, she could be found at the MIT Senseable City Lab, and her design work was featured on the cover of *Nature*. Previously, she studied architecture and mathematics at Harvard University. Snoweria also draws a webcomic called *Lonesome Whales*.

INDEX